Math Made Nice-n-Easy Books™

In This Book:

- ## Tangents, Normals, & Slopes of Curves

- ## Limits & Differentiation

- ## Derivatives

- ## Integration

"MATH MADE NICE-n-EASY #8" is one in a series of books designed to make the learning of math interesting and fun. For help with additional math topics, see the complete series of "MATH MADE NICE-n-EASY" titles.

Based on U.S. Government Teaching Materials

Research & Education Association
61 Ethel Road West
Piscataway, New Jersey 08854
Dr. M. Fogiel, Director

MATH MADE NICE-N-EASY BOOKS™
BOOK #8

Copyright © 2002 by Research & Education Association. This copyright does not apply to the information included from U.S. Government publications, which was edited by Research & Education Association.

Printed in the United States of America

Library of Congress Control Number 2001091345

International Standard Book Number 0-87891-207-X

MATH MADE NICE-N-EASY is a trademark of Research & Education Association, Piscataway, New Jersey 08854

WHAT "MATH MADE NICE-N-EASY" WILL DO FOR YOU

The "Math Made Nice-n-Easy" series simplifies the learning and use of math and lets you see that math is actually interesting and fun. This series of books is for people who have found math scary, but who nevertheless need some understanding of math without having to deal with the complexities found in most math textbooks.

The "Math Made Nice-n-Easy" series of books is useful for students and everyone who needs to acquire a basic understanding of one or more math topics. For this purpose, the series is divided into a number of books which deal with math in an easy-to-follow sequence beginning with basic arithmetic, and extending through pre-algebra, algebra, and calculus. Each topic is described in a way that makes learning and understanding easy.

Almost everyone needs to know at least some math at work, or in a course of study.

For example, almost all college entrance tests and professional exams require solving math problems. Also, almost all occupations (waiters, sales clerks, office people) and all crafts (carpentry, plumbing, electrical) require some ability in math problem solving.

The "Math Made Nice-n-Easy" series helps the reader grasp quickly the fundamentals that are needed in using math. The reader is led by the hand, step-by-step, through the various concepts and how they are used.

By acquiring the ability to use math, the reader is encouraged to further his/her skills and to forget about any initial math fears.

The "Math Made Nice-n-Easy" series includes material originated by U.S. Government research and educational efforts. The research was aimed at devising tutoring and teaching methods for educating government personnel lacking a technical and/or mathematical background. Thanks for these efforts are due to the U.S. Bureau of Naval Personnel Training.

Dr. Max Fogiel
Program Director

Contents

Chapter 11
Tangents, Normals, and Slopes of Curves 1
Slope of a Curve at a Point.. 1
Direction of a Curve 3
Tangent at a Given Point on the Standard Parabola.................... 6
Equations of Tangents and Normals 19
Lengths of Subtangent and Subnormals.................................. 23
Parametric Equations 29
Motion in a Straight Line 29
Motion in a Circle 31
Other Parametric Equations 34

Chapter 12
Limits and Differentiation 45
Limit Concept 45
Definition of Limit 45
Indeterminate Forms 55
Limit Formulas 58
Infinitesimals 62
Definitions 63
Products 65
Conclusions 67
Discontinuities 67
Increments and Differentiations 73
General Formula 77
Examples of Differentiation 78

Chapter 13
Derivatives

Derivatives 89
Derivative of Constant 90
Formula 90
Proof 90
Variables 93
Power Form 93
Sums 98
Products 102
Quotients 105
Powers of Functions 107
Radicals 109
Chain Rules 114
Inverse Functions 117
Implicit Functions 119
Trigonometric Problems 122

Chapter 14
Integration

Integration 128
Definitions 128
Interpretation of an Integral 129
Area Under a Curve 129
Constant of Integration 140
Integrand 140
Indefinite Integrals 141
Evaluating the Constant 143
Rules for Integration 147
Definite Integrals 154
Upper and Lower Limits 155

Chapter 11
Tangents, Normals, & Slopes of Curves

In chapter 9, the notation $\frac{\Delta y}{\Delta x}$ was introduced to represent the slope of a line. The straight line discussed has a constant slope and the symbol Δy was defined as $(y_2 - y_1)$ and Δx was defined as $(x_2 - x_1)$. In this chapter we will discuss the slope of curves at specific points on the curves. We will do this with as little calculus as possible but our discussion will be directed toward the study of calculus.

Slope of a Curve at a Point

In figure 11-1, the slope of the curve is represented at two different places by $\frac{\Delta y}{\Delta x}$. The value of $\frac{\Delta y}{\Delta x}$ taken on the lower part of the curve will be extremely close to the actual slope at P_1 because P_1 lies on a nearly straight portion of the curve. The value of the slope at P_2 will be less accurate than the slope near P_1 because P_2 lies on a portion of the curve which has more curvature than the portion of the curve

1

near P_1. In order to obtain an accurate measure of the slope of the curve at each point, as small a portion of the curve as possible should be used. When the curve is nearly a straight line a very small error will occur when finding the slope regardless of the value of the increments Δy and Δx. If the curvature is great and large increments are used when finding the slope of a curve, the error will become very large.

Thus, it follows that the error can be reduced to an infinitesimal if the increments are chosen infinitely small. Whenever the slope of a curve at a given point is desired, the increments Δy and Δx should be extremely small. Consequently, the arc of the curve can be replaced by a straight line, which determines the slope of the tangent at that point.

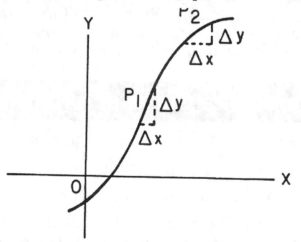

Figure 11-1.—Curve with increments Δy and Δx.

It must be understood that when we speak of the tangent to a curve at a specific point we are really considering the secant line, which cuts a curve in at least two points. This secant line is to be decreased in length, keeping the end points on the curve, to such a small value that it may be considered to be a point. This point is then extended to form the tangent to the curve at that specific point. Figure 11-2 shows this concept.

Direction of a Curve

If we allow

$$y = f(x)$$

to represent the equation of a curve, then $\frac{\Delta y}{\Delta x}$ is the slope of the line tangent to the curve at P (x, y).

The direction of a curve is defined as the direction of the tangent line at any point on the curve. Let θ equal the inclination of the tangent line; then the slope equals $\tan \theta$ and

$$\frac{\Delta y}{\Delta x} = \tan \theta$$

is the slope of the curve at any point P (x, y). The angle θ_1 is the inclination of the tangent to the curve at P_1 in figure 11-3. This angle is acute and the value of $\tan \theta_1$ is positive. Hence the slope is positive at point P_1. The angle θ_2 is an obtuse angle and $\tan \theta_2$ is negative and the slope at point P_2 is negative. All lines which lean to the right have positive

3

slopes and all lines which lean to the left have negative slopes. At point P_3 the tangent to the curve is horizontal and θ equals 0. This means that

$$\frac{\Delta y}{\Delta x} = \tan 0° = 0$$

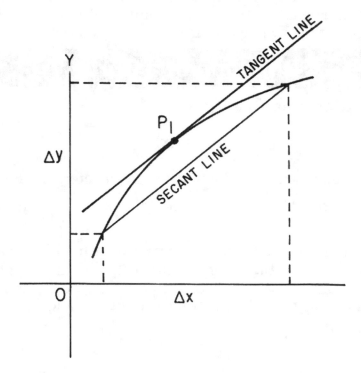

Figure 11-2.—Curve with secant line and tangent line.

The fact that the slope of a curve is zero when the tangent to the curve at that point is horizontal is of great importance in calculus when determining the maximum or minimum points of a curve. Whenever the slope of a curve is zero, the curve may be at either a maximum or a minimum.

Whenever the inclination of the tangent to a curve at a point is 90°, the tangent line is

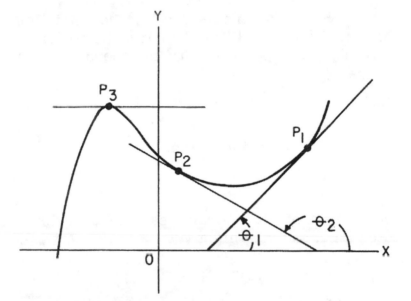

Figure 11-3.—Curve with tangent lines.

vertical and parallel to the Y axis. This results in an infinitely large slope

$$\frac{\Delta y}{\Delta x} = \tan 90° = \infty$$

5

Tangent at a Given Point on the Standard Parabola

The standard parabola is represented by the equation

$$y^2 = 4ax$$

Let P_1 with coordinates (x_1, y_1) be a point on the curve. Choose P' on the curve, figure 11-4, near the given point so that the coordinates of P' are

$$(x_1 + \Delta x, \; y_1 + \Delta y)$$

Since P' is a point on the curve

$$y^2 = 4ax$$

the values of its coordinates may be substituted for x and y. This gives

$$(y_1 + \Delta y)^2 = 4a(x_1 + \Delta x)$$

or

$$y_1^2 + 2y_1 \, \Delta y + (\Delta y)^2 = 4ax_1 + 4a \, \Delta x \qquad (1)$$

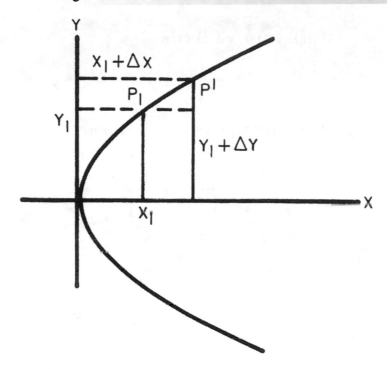

Figure 11-4.—Parabola.

The point $P_1(x_1, y_1)$ also lies on the curve and we have

$$y_1^2 = 4ax_1$$

Substituting this value for y_1^2 into equation (1) transforms it into

$$4ax_1 + 2y_1 \, \Delta y + (\Delta y)^2 = 4ax_1 + 4a \, \Delta x$$

Simplifying we obtain

$$2y_1 \Delta y + (\Delta y)^2 = 4a\Delta x$$

Divide through by Δx, obtaining

$$\frac{2y_1 \Delta y}{\Delta x} + \frac{(\Delta y)^2}{\Delta x} = \frac{4a\Delta x}{\Delta x}$$

which gives

$$2y_1 \frac{\Delta y}{\Delta x} = 4a - \frac{(\Delta y)^2}{\Delta x}$$

Solving for $\frac{\Delta y}{\Delta x}$ we find

$$\frac{\Delta y}{\Delta x} = \frac{4a}{2y_1} - \frac{\frac{(\Delta y)^2}{\Delta x}}{2y_1}$$

$$= \frac{2a}{y_1} - \frac{\frac{(\Delta y)^2}{\Delta x}}{2y_1}$$

(3)

Before proceeding, a discussion of the term

in equation (3) is in order. If we solve equation (2) for Δy we find

$$2y_1 \, \Delta y + (\Delta y)^2 = 4a\Delta x$$

then

$$\Delta y(2y_1 + \Delta y) - 4a\Delta x$$

and

$$\Delta y = \frac{4a\Delta x}{2y_1 + \Delta y}$$

Since the denominator contains a term not dependent upon Δy or Δx, as we let Δx approach zero Δy will also approach zero.

NOTE: We may find a value for Δx that will make Δy less than 1 and then when Δy is squared it will approach zero at least as rapidly as Δx does.

We now refer to equation (3) again and make the statement that we may disregard $\dfrac{\frac{(\Delta y)^2}{\Delta x}}{2y_1}$ since it approaches zero when Δx approaches zero.

Then

$$\frac{\Delta y}{\Delta x} = \frac{2a}{y_1}$$

The quantity $\frac{\Delta y}{\Delta x}$ is the slope of the line connecting P_1 and P'. From figure 11-4, it is obvious that the slope of the curve at P_1 is different from the slope of the line connecting P_1 and P'.

As Δx and Δy approach zero, the ratio $\frac{\Delta y}{\Delta x}$ will approach more and more closely the true slope of the curve at P_1. We designate the slope by (m). Thus, as Δx approaches zero, equation (4) becomes

$$m = \frac{2a}{y_1}$$

The equation for a straight line in the point slope form is

$$y - y_1 = m(x - x_1)$$

Substituting $\frac{2a}{y_1}$ for m gives

$$y - y_1 = \frac{2a}{y_1}(x - x_1)$$

Clearing fractions we have

$$yy_1 - y_1^2 = 2ax - 2ax_1 \qquad (5)$$

but

$$y_1^2 = 4ax_1 \qquad (6)$$

Adding equations (5) and (6) yields

$$yy_1 = 2ax + 2ax_1$$

Dividing by y_1 gives

$$y = \frac{2ax}{y_1} + \frac{2ax_1}{y_1}$$

which is an equation of a straight line in the slope intercept form. This is the equation of the tangent line to the parabola

$$y^2 = 4ax$$

at the point (x_1, y_1).

EXAMPLE: Given the equation

$$y^2 = 8x$$

find the slope of the curve and the equation of the tangent line at the point (2, 4).

SOLUTION: Put the equation in standard form as follows: Solve for (a) by letting

$$y^2 = 8x$$

have the form

$$y^2 = 4ax$$

Then

$$4a = 8$$

$$a = 2$$

and

$$2a = 4$$

The slope m at point (2, 4) becomes

$$m = \frac{2a}{y_1}$$

$$= \frac{2(2)}{4} = 1$$

The slope of the line is 1 and the equation of the tangent to the curve at the point (2, 4) is

$$y = \frac{2ax}{y_1} + \frac{2ax_1}{y_1}$$

$$= \frac{(2)\ (2)\ (x)}{4} + \frac{(2)\ (2)\ (2)}{4}$$

$$= x + 2$$

This method, used to find the slope and equation of the tangent for a standard parabola, can be used to find the slope and equation of the tangent to a curve at any point regardless of the type of curve. The method can be used to find these relationships for circles, hyperbolas, ellipses, and general algebraic curves.

This general method is outlined as follows: To find the slope (m) of a given curve at the point $P_1 (x_1, y_1)$ choose a second point P' on the curve so that it has coordinates $(x_1 + \Delta x, y_1 + \Delta y)$ and substitute the coordinates of P' in the equation of the curve and simplify. Divide through by Δx and eliminate terms which contain powers of Δy higher than the first power, as previously discussed. Let Δx approach zero and $\frac{\Delta y}{\Delta x}$ will approach the absolute value for the slope (m) at point P_1. Finally solve for (m).

When the slope and coordinates of a point on the curve are known, the equation of the tangent line can be found by using the point slope method.

EXAMPLE: Using the method outlined, find the slope and equation of the tangent line to the curve

$$x^2 + y^2 = r^2 \text{ at } (x_1, y_1) \qquad (7)$$

SOLUTION: Choose a second point P_1 such that it has coordinates

$$(x_1 + \Delta x, y_1 + \Delta y)$$

Substitute into equation (7) and

$$(x_1 + \Delta x)^2 + (y_1 + \Delta y)^2 = r^2$$

thus

$$x_1^2 + 2x_1\Delta x + (\Delta x)^2 + y_1^2 + 2y_1\Delta y + (\Delta y)^2 = r^2$$

then

$$2x_1\Delta x + (\Delta x)^2 + 2y_1\Delta y + (\Delta y)^2 = r^2 - x_1^2 - y_1^2$$

$$= r^2 - (x_1^2 + y_1^2)$$

$$= 0$$

Divide through by Δx

$$\frac{2x_1 \Delta x}{\Delta x} + \frac{(\Delta x)^2}{\Delta x} + \frac{2y_1 \Delta y}{\Delta x} + \frac{(\Delta y)^2}{\Delta x} = 0$$

and eliminating $(\Delta y)^2$ results in

$$2x_1 + \Delta x + 2y_1 \frac{\Delta y}{\Delta x} = 0$$

Solve for $\frac{\Delta y}{\Delta x}$ as follows:

$$\frac{\Delta y}{\Delta x} = \frac{-2x_1 - \Delta x}{2y_1}$$

but

$$\frac{\Delta y}{\Delta x} = m$$

and

$$m = \frac{-2x_1 - \Delta x}{2y_1}$$

Let Δx approach zero and

$$m = \frac{-x_1}{y_1}$$

Now use the point slope form of a straight line with the slope equal to $\dfrac{-x_1}{y_1}$ and find at point (x_1, y_1)

$$y - y_1 = m(x - x_1)$$

$$= \frac{-x_1}{y_1}(x - x_1)$$

Rearranging

$$yy_1 - y_1^2 = -x_1 x + x_1^2$$

and

$$yy_1 = -x_1 x + x_1^2 + y_1^2$$

but

$$x_1^2 + y_1^2 = r^2$$

Then, by substitution

$$yy_1 = -x_1 x + r^2$$

and

$$y = \frac{-x_1 x}{y_1} + \frac{r^2}{y_1}$$

which is the general equation of the tangent line to the curve

$$x^2 + y^2 = r^2 \text{ at } (x_1, y_1)$$

EXAMPLE: Using the given method, with minor changes, find the slope and equation of the tangent line to the curve

$$x^2 - y^2 = k^2 \text{ at } (x_1, y_1) \tag{8}$$

SOLUTION: Choose a second point P_1 such that it has coordinates

$$(x_1 + \Delta x, y_1 + \Delta y)$$

Substitute into equation (8) and

$$(x_1 + \Delta x)^2 - (y_1 + \Delta y)^2 = k^2$$

and

$$x_1^2 + 2x_1 \Delta x + (\Delta x)^2 - y_1^2 - 2y_1 \Delta y - (\Delta y)^2 = k^2 \tag{9}$$

then subtract equation (8) from (9) and obtain

$$2x_1 \Delta x + (\Delta x)^2 - 2y_1 \Delta y - (\Delta y)^2 = 0$$

then divide by Δx and we have

$$2x_1 + \Delta x - 2y_1 \frac{\Delta y}{\Delta x} - \frac{(\Delta y)^2}{\Delta x} = 0$$

Let Δx approach zero and

$$2x_1 - 2y_1 \frac{\Delta y}{\Delta x} = 0$$

Solving for $\frac{\Delta y}{\Delta x}$ results in

$$\frac{\Delta y}{\Delta x} = m = \frac{x_1}{y_1}$$

Use the point slope form of a straight line to find the equation of the tangent line at point (x_1, y_1) as shown in the following:

$$y - y_1 = m(x - x_1)$$

Substitute $\frac{x_1}{y_1}$ for m

$$y - y_1 = \frac{x_1}{y_1}(x - x_1)$$

Multiply through by y_1 and

$$yy_1 - y_1^2 = x_1 x - x_1^2$$

Rearrange to obtain

$$yy_1 = x_1 x - x_1^2 + y_1^2$$

$$= x_1 x - \left(x_1^2 - y_1^2\right)$$

Substitute $x_1^2 - y_1^2$ for k^2 and

$$yy_1 = x_1 x - k^2$$

Divide through by y_1 to obtain

$$y = \frac{x_1 x}{y_1} - \frac{k^2}{y_1}$$

which is the equation and slope desired.

Practice Problems

Fined the slope (m) and equaltion of the tangent line, in problems1 through 6, at the given point.

1. $y^2 = \frac{4}{3}x$ at (3, 2)

2. $y^2 = 12x$ at (3, 6)

3. $x^2 + y^2 = 25$ at (-3, 4)

4. $x^2 + y^2 = 100$ at (6, 8)

5. $x^2 - y^2 = 9$ at (5, 4)

6. $x^2 - y^2 = 3$ at (2, 1)

7. Find the slope of $y = x^2$ at (2, 4)

8. Find the slope of
   $y = 2x^2 - 3x + 2$ at (2, 4)

Answers

1. $y = \frac{x}{3} + 1$, $m = \frac{1}{3}$

2. $y = x + 3$, $m = 1$

3. $y = \frac{3x}{4} + \frac{25}{4}$, $m = \frac{3}{4}$

4. $y = \dfrac{-3x}{4} + \dfrac{25}{2}$, $m = \dfrac{-3}{4}$

5. $y = \dfrac{5x}{4} - \dfrac{9}{4}$, $m = \dfrac{5}{4}$

6. $y = 2x - 3$, $m = 2$

7. $m = 4$

8. $m = 5$

Equations of Tangents and Normals

In figure 11-5, the coordinates of point P_1 on the curve are (x_1, y_1). Let the slope of the tangent to the curve at point P_1 be denoted by m_1. Knowing the slope and a point through which the tangent line passes, the equation of that tangent line can be determined by using the point slope form.

Thus, the equation of the tangent line (MP_1) is

$$y - y_1 = m(x - x_1)$$

The normal to a curve at a point (x_1, y_1) is the line which is perpendicular to the tangent line at that point. The slope of the normal line is then $-\dfrac{1}{m_1}$ where, as before, the slope of the tangent line is m_1. This is shown in the following:

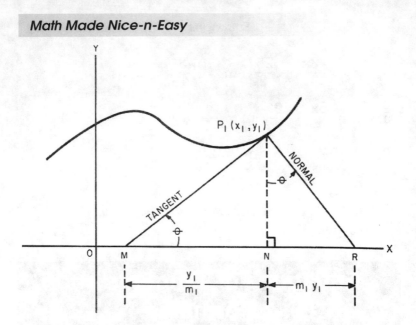

Figure 11-5.—Curve with tangent and
normal lines.

If

$$m_1 = \tan \theta$$

then

$$m_2 = \tan (\theta + 90°)$$

$$= -\tan [180° - (\theta + 90°)]$$

$$= -\tan (90° - \theta)$$

$$= -\cot \theta$$

$$= -\frac{1}{\tan \theta}$$

$$= - \frac{1}{m_1}$$

therefore

$$m_2 = - \frac{1}{m_1}$$

The equation of the normal through P_1 is

$$y - y_1 = - \frac{1}{m_1}(x - x_1)$$

Notice that if the slope of the tangent is m_1 and the slope of the normal to the tangent is m_2 and

$$m_2 = - \frac{1}{m_1}$$

then the product of the slopes of the tangent and normal equals -1. The relationship between the slopes of the tangent and normal stated more formally is: The slope of the normal is the negative reciprocal of the slope of the tangent.

Another approach to show the relationship between the slopes of the tangent and normal follows: The inclination of one line must be 90° greater than the other. Then

$$\theta_2 = \theta_1 + 90°$$

If

$$\tan \theta_2 = m_2$$

and

$$\tan \theta_2 = \tan (\theta_1 + 90°)$$

then

$$\tan (\theta_1 + 90°) = \frac{\sin (\theta_1 + 90°)}{\cos (\theta_1 + 90°)}$$

$$= \frac{\sin \theta_1 \cos 90° + \cos \theta_1 \sin 90°}{\cos \theta_1 \cos 90° - \sin \theta_1 \sin 90°}$$

$$= - \frac{\cos \theta_1}{\sin \theta_1}$$

$$= - \cot \theta_1$$

$$= - \frac{1}{\tan \theta_1}$$

therefore

$$\tan \theta_2 = - \frac{1}{\tan \theta_1}$$

Lengths of Subtangent and Subnormals

The length of the tangent is defined as that portion of the tangent line between the point $P_1 (x_1, y_1)$ and the point where the tangent line crosses the X axis. In figure 11-5, the length of the tangent is (MP_1).

The length of the normal is defined as that portion of the normal line between the point P_1 and the X axis. That is $(P_1 R)$ which is perpendicular to the tangent.

The projections of these lines on the X axis are known as the length of the subtangent (MN) and the length of the subnormal (NR).

The relationships between the slope of the tangent and the lengths of the subtangent and subnormal follows:

From the triangle $MP_1 N$, in figure 11-5,

$$\tan \theta = m_1 = \frac{P_1 N}{MN}$$

and

$$MN = \frac{P_1 N}{m_1}$$

23

The line segment (MN) is the length of the subtangent and (P_1N) is equal to the vertical coordinate y_1. Therefore, the length of the subtangent is $\dfrac{y_1}{m_1}$.

In the triangle NP_1R,

$$\tan \theta = \frac{NR}{NP_1}$$

but

$$\tan \theta = m_1$$

and

$$NR = m_1 NP_1$$
$$= m_1 y_1$$

Therefore, the length of the subnormal is $m_1 y_1$.

From this, as shown in figure 11-5, the length of the tangent and normal may be found by using the Pythagorean theorem.

NOTE: If the subtangent lies to the right of point M, it is considered positive, if to the left it is negative. Likewise, if the subnormal extends to the right of N it is positive, to the left it is negative.

EXAMPLE: Find the equation of the tangent, the equation of the normal, the length of the subtangent, the length of the subnormal, and the lengths of the tangent and normal of

24

$$y^2 = \frac{4}{3}x, \text{ at } (3, 2)$$

SOLUTION: Find the value of 2a from

$$y^2 = 4ax$$

Since

$$y^2 = \frac{4}{3}x$$

then

$$4a = \frac{4}{3}$$

$$a = \frac{1}{3}$$

$$2a = \frac{2}{3}$$

The slope is

$$m = \frac{2a}{y_1} = \frac{\frac{2}{3}}{2} = \frac{1}{3}$$

Using the point slope form

$$y - y_1 = m(x - x_1)$$

then, at point (3, 2)

$$y - 2 = \frac{1}{3}(x - 3)$$

$$= \frac{x}{3} - 1$$

and

$$y = \frac{x}{3} + 1$$

which is the equation of the tangent line.

Use the negative reciprocal of the slope to find the equation of the normal as follows:

$$y - 2 = -3(x - 3)$$

$$= -3x + 9$$

then

$$y = -3x + 11$$

The length of the subtangent is

$$\frac{y_1}{m_1} = \frac{2}{\frac{1}{3}} = 6$$

and the length of the subnormal is

$$y_1 m_1 = 2\left(\frac{1}{3}\right) = \frac{2}{3}$$

To find the length of the tangent we use the Pythagorean theorem. Thus, the length of the tangent is

$$\sqrt{\left(\frac{y_1}{m_1}\right)^2 + (y_1)^2}$$

$$= \sqrt{(6)^2 + (2)^2}$$

$$= \sqrt{40}$$

$$= 6.32$$

The length of the normal is equal to

$$\sqrt{(y_1 m_1)^2 + (y_1)^2}$$

$$= \sqrt{\left(\frac{2}{3}\right)^2 + (2)^2}$$

$$= \sqrt{\frac{40}{9}}$$

$$= \frac{6.32}{3}$$

Practice Problems

Find the equation of the tangent and normal, and the lengths of the subtangent and subnormal in the following:

1. $y^2 = 12x$,   at $(3, 6)$

2. $x^2 + y^2 = 25$,   at $(-3, 4)$

3. $x^2 - y^2 = 9$,   at $(5, 4)$

4. $y = 2x^2 - 3x + 2$,   at $(1, 1)$

Answers

1. Equation of tangent $y = x + 3$

     Equation of normal $y = -x + 9$

     Length of subtangent   6

     Length of subnormal   6

2. Equation of tangent $y = \dfrac{3x}{4} + \dfrac{25}{4}$

     Equation of normal $y = \dfrac{-4x}{3}$

     Length of subtangent $\dfrac{16}{3}$

     Length of subnormal   5

3. Equation of tangent $y = \dfrac{5x}{4} - \dfrac{9}{4}$

     Equation of normal $y = \dfrac{-4x}{5} + 8$

     Length of subtangent $\dfrac{5}{16}$

     Length of subnormal   5

4. Equation of tangent $y = x$

   Equation of normal $y = -x + 2$

   Length of subtangent   1

   Length of subnormal   1

Parametric Equations

Equations used previously have been functions involving two unknowns such as x and y. The functions have been in either Cartesian or polar coordinates and have been defined by one equation. If a third variable is introduced it is called a parameter. When two equations are used, each containing the parameter, the equations are called parametric equations.

Motion in a Straight Line

To illustrate the application of a parameter we will assume that an aircraft takes off from a field which we will call the origin. Figure 11-6 shows the diagram we will use. The aircraft is flying on a compass heading of due north. There is a wind blowing from the west at 20 miles per hour and the airspeed of the aircraft is 400 miles per hour. Let the direction of the positive Y axis be due north and the positive X axis be due east as shown in figure 11-6. Use the scales as shown.

29

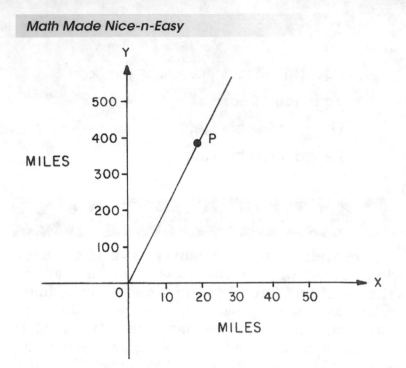

Figure 11-6.—Aircraft position.

One hour after takeoff the position of the aircraft, represented by point P, is 400 miles north and 20 miles east of the origin. If we use t as the parameter, then at any time t the aircrafts position (x, y) will be given by x equals 20t and y equals 400t.
The equations are

$$x = 20t$$

and

$$y = 400\ t$$

and are called parametric equations. Notice that time is not plotted on the graph of figure 11-6. The parameter t is used only to plot the position (x, y) of the aircraft.

We may eliminate the parameter t to obtain a direct relationship between x and y as follows:

If

$$t = \frac{x}{20}$$

then

$$y = 400 \left(\frac{x}{20}\right)$$

$$y = 20x$$

and we find the graph to be a straight line. When we eliminated the parameter the result was the rectangular coordinate equation of the line.

Motion in a Circle

Consider the parametric equations

$$x = r \cos t$$

and

$$y = r \sin t$$

These equations describe the position of a point (x, y) at any time t. They can be transposed into a single equation by squaring both sides of each equation to obtain

31

$$x^2 = r^2 \cos^2 t$$

$$y^2 = r^2 \sin^2 t$$

and adding

$$x^2 + y^2 = r^2 \cos^2 t + r^2 \sin^2 t$$

Rearranging we have

$$x^2 + y^2 = r^2 (\cos^2 t + \sin^2 t)$$

but

$$\cos^2 t + \sin^2 t = 1$$

then

$$x^2 + y^2 = r^2$$

which is the equation of a circle.

This means that if various values were assigned to t and the corresponding values of x and y were calculated and plotted, the result would be a circle. In other words, the point (x, y) moves in a circular path.

Using this example again, that is

$$x = r \cos t$$

and

$$y = r \sin t$$

and given that

$$m_2 = \frac{\Delta x}{\Delta t} = -r \sin t$$

and

$$m_1 = \frac{\Delta y}{\Delta t} = r \cos t$$

we are able to express the slope at any point on the circle in terms of t.

NOTE: These expressions for $\frac{\Delta x}{\Delta t}$ and $\frac{\Delta y}{\Delta t}$ may be found by using calculus, but we will accept them for the present without proof.

If we know $\frac{\Delta y}{\Delta t}$ and $\frac{\Delta x}{\Delta t}$, we may find $\frac{\Delta y}{\Delta x}$ which is the slope of a curve at any point.

That is,

$$m = \frac{\Delta y}{\Delta x} = \frac{\frac{\Delta y}{\Delta t}}{\frac{\Delta x}{\Delta t}}$$

By substituting we find

$$\frac{\Delta y}{\Delta x} = \frac{r \cos t}{-r \sin t} = -\cot t$$

Comparing this result with equation (7) of the previous section, we find that in rectangular coordinates the slope is given as

$$m = -\frac{x_1}{y_1}$$

33

while in terms of a parameter it is

$$m = - \cot t$$

Other Parametric Equations

EXAMPLE: Find the equation of the tangent and the normal and the length of the subtangent and the subnormal for the curve represented by

$$x = t^2$$

and

$$y = 2t + 1$$

at

$$t = 1$$

given that

$$\frac{\Delta x}{\Delta t} = 2t$$

and

$$\frac{\Delta y}{\Delta t} = 2$$

SOLUTION: Since t equals 1 we write

$$x = 1$$

and

$$y = 3$$

and

$$\frac{\Delta y}{\Delta x} = \frac{2}{2t} = \frac{1}{t}$$

then

$$m = \frac{1}{t}$$

The equation of the tangent line when t is equal to 1 is

$$y - 3 = 1(x - 1)$$

$$y = x + 2$$

The equation of the normal line is

$$y - 3 = -1(x - 1)$$

$$y = -x + 4$$

The length of the subtangent is

$$\frac{y_1}{m} = \frac{3}{1} = 3$$

The length of the subnormal

$$y_1 m = (1)(3) = 3$$

EXAMPLE: Find the equation of the tangent and normal and the lengths of the subtangent and subnormal to the curve represented by the parametric equations

35

$$x = 2 \cos \theta$$

and

$$y = 2 \sin \theta$$

at the point where

$$\theta = 45°$$

given that

$$\frac{\Delta x}{\Delta \theta} = -2 \sin \theta$$

and

$$\frac{\Delta y}{\Delta \theta} = 2 \cos \theta$$

SOLUTION: We know that

$$\frac{\Delta y}{\Delta x} = \frac{\dfrac{\Delta y}{\Delta \theta}}{\dfrac{\Delta x}{\Delta \theta}} = \frac{2 \cos \theta}{-2 \sin \theta} = - \cot \theta$$

Then at the point where

$$\theta = 45°$$

we have

$$m = \frac{\Delta y}{\Delta x} = - \cot 45° = -1$$

In order to find (x_1, y_1) we substitute

$$\theta = 45°$$

in the parametric equations and

$$x_1 = 2 \cos 45° = 2 \left(\frac{\sqrt{2}}{2}\right) = \sqrt{2}$$

$$y_1 = 2 \sin 45° = 2 \left(\frac{\sqrt{2}}{2}\right) = \sqrt{2}$$

The equation of the tangent is

$$y - y_1 = m(x - x_1)$$

Substituting we have

$$y - \sqrt{2} = -1(x - \sqrt{2})$$

or

$$x + y = 2\sqrt{2}$$

The equation of the normal is

$$y - \sqrt{2} = 1(x - \sqrt{2})$$

or

$$x - y = 0$$

The length of the subtangent is

$$\frac{y_1}{m} = \frac{\sqrt{2}}{-1} = -\sqrt{2}$$

The length of the subnormal is

$$y_1 m = (\sqrt{2})(-1) = -\sqrt{2}$$

The horizontal and vertical tangents of a curve can be found very easily when the curve is represented by parametric equations. The slope of a curve at any point equals zero when the tangent is parallel to the x axis. In parametric equations the horizontal and vertical tangents can be found easily by setting

$$\frac{\Delta y}{\Delta t} = 0$$

and

$$\frac{\Delta x}{\Delta t} = 0$$

For the horizontal tangent solve $\frac{\Delta y}{\Delta t}$ equals zero for t and for the vertical tangent solve $\frac{\Delta x}{\Delta t}$ equals zero for t.

EXAMPLE: Find the points of contact of the horizontal and vertical tangents to the curve represented by the parametric equations

$$x = 3 - 4 \sin \theta$$

and

$$y = 4 + 3 \cos \theta$$

Plot the graph of the function by taking θ from 0° to 360° in increments of 30°.
Given that

$$\frac{\Delta x}{\Delta \theta} = - 4 \cos \theta$$

and

$$\frac{\Delta y}{\Delta \theta} = - 3 \sin \theta$$

38

SOLUTION: The graph of the function shows that the figure is an ellipse, figure 11-7, and consequently there will be two horizontal and two vertical tangents. The coordinates of the horizontal tangent points are found by first setting

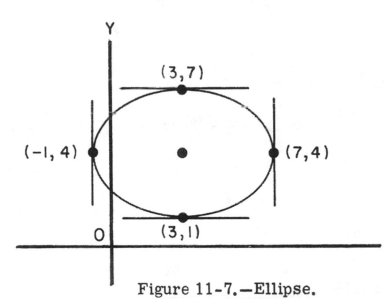

Figure 11-7.—Ellipse.

$$\frac{\Delta y}{\Delta \theta} = 0$$

This gives

$$-3 \sin \theta = 0$$

Then

$$\sin \theta = 0$$

and

$$\theta = 0° \text{ or } 180°$$

Substituting 0° we have

$$x = 3 - 4 \sin 0°$$

$$= 3 - 0$$

$$= 3$$

and

$$y = 4 + 3 \cos 0°$$

$$= 4 + 3$$

$$= 7$$

Substituting 180° we obtain

$$x = 3 - 4 \sin 180°$$

$$= 3 - 0$$

$$= 3$$

and

$$y = 4 + 3 \cos 180°$$

$$= 4 - 3$$

$$= 1$$

The coordinates of the points of contact of the horizontal tangents to the ellipse are (3, 1) and (3, 7).

The coordinates of the vertical tangent points of contact are found by setting

$$\frac{\Delta y}{\Delta \theta} = 0$$

We find

$$- 4 \cos \theta = 0$$

from which

$$\theta = 90° \text{ or } 270°$$

Substituting 90° we obtain

$$x = 3 - 4 \sin 90°$$

$$= 3 - 4$$

$$= -1$$

and

$$y = 4 + 3 \cos 90°$$

$$= 4 + 0$$

$$= 4$$

Substituting 270° gives

$$x = 3 - 4 \sin 270°$$

41

$$= 3 + 4$$

$$= 7$$

and

$$y = 4 + 3 \cos 270°$$

$$= 4 + 0$$

$$= 4$$

The coordinates of the points of contact of the vertical tangents to the ellipse are (-1, 4) and (7, 4).

Practice Problems

Find the equations of the tangent and the normal and the lengths of the subtangent and the subnormal for each of the following curves at the point indicated.

1. $x = t^3$

   $y = 3t$

   at $t = -1$

   given $\dfrac{\Delta x}{\Delta t} = 3t^2$ and $\dfrac{\Delta y}{\Delta t} = 3$

2. $x = t^2 + 8$

   $y = t^2 + 1$

   at $t = 2$

   given $\dfrac{\Delta x}{\Delta t} = 2t$ and $\dfrac{\Delta y}{\Delta t} = 2t$

3. $x = t$

   $y = t^2$

   at $t = 1$

   given $\dfrac{\Delta x}{\Delta t} = 1$ and $\dfrac{\Delta y}{\Delta t} = 2t$

4. Find the points of contact of the horizontal and vertical tangents to the curve

$$x = 2 \cos \theta$$

$$y = 3 \sin \theta$$

given

$$\frac{\Delta x}{\Delta \theta} = -2 \sin \theta$$

$$\frac{\Delta y}{\Delta \theta} = 3 \cos \theta$$

Answers

1. Equation of tangent $y = x - 2$
   Equation of normal $y = -x - 4$
   Length of subtangent   3
   Length of subnormal   3

2. Equation of tangent $y = x - 7$
   Equation of normal $y = -x + 17$
   Length of subtangent   5
   Length of subnormal   5

3. Equation of tangent $y = 2x - 1$

   Equation of normal $y = \dfrac{-x}{2} + \dfrac{3}{2}$

   Length of subtangent $\dfrac{1}{2}$

   Length of subnormal 2

4. Coordinates of the points of contact of the horizontal tangent to the ellipse are (0, 3) and (0, -3) and the vertical tangent to the ellipse are (2, 0) and (-2, 0).

Chapter 12
Limits & Differentiation

Limits and differentiation are the beginning of the study of calculus, which is an important and powerful method of computation.

Limit Concept

The study of the limit concept is very important as it is the very heart of the theory and operation of calculus. We will include in this section the definition of limit, some of the indeterminate forms of limits, and some limit formulas, along with example problems.

Definition of Limit

Before we start differentiation there are certain concepts which we must understand. One of these concepts deals with the limit of a function. Many times it will be necessary to find the value of the limit of a function.

The discussion of limits will begin with an intuitive point of view.

We will work with the equation

$$y = f(x) = x^2$$

which is shown in figure 12-1. The point P represents the point corresponding to

$$y = 16$$

and

$$x = 4$$

The behavior of y for given values of x near the point

$$x = 4$$

is the center of the discussion. For the present we will exclude the point P which is encircled on the graph.

We will start with values lying between

$$x = 2$$

and

$$x = 6$$

indicated by line A in figure 12-1. This interval may be written as

$$0 < |x - 4| < 2$$

The corresponding interval for y is between

$$y = 4$$

and

$$y = 36$$

We now take a smaller interval about

$$x = 4 \text{ (line B)}$$

by using values of

$$x = 3$$

and

$$x = 5$$

and find the corresponding interval for y to be between

$$y = 9$$

and

$$y = 25$$

Interval of x		Interval of f(x)
2 – 6	(A)	4 – 36
3 – 5	(B)	9 – 25
3.5 – 4.5	(C)	12.25 – 20.25
3.9 – 4.1	(D)	15.21 – 16.81

(A)

47

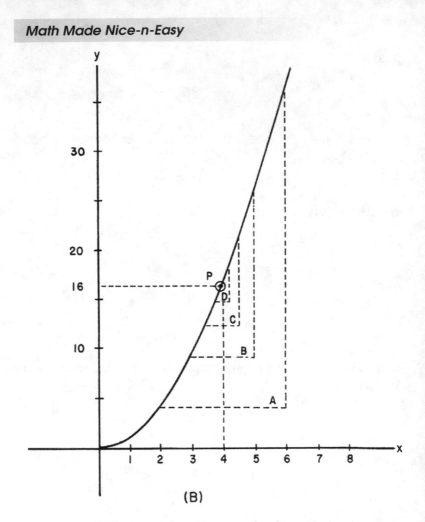

Figure 12-1.—(A) value chart.
(B) Graph of $y = x^2$;

These intervals for x and y are written as

$$0 < |x - 4| < 1$$

and

$$9 < y < 25$$

As we diminish the interval of x around

$$x = 4 \text{ (line C and line D)}$$

we find the values of

$$y = x^2$$

to be grouped more and more closely around

$$y = 16$$

This is shown by the chart in figure 12-1.

Although we have used only a few intervals of x in the discussion, it should be apparent that we can make the values about y group as closely as we desire by merely limiting the values assigned to x about

$$x = 4$$

Because the foregoing is true, we may now say that the limit of x^2, as x approaches 4, results in the value 16 for y and we write

$$\lim_{x \to 4} x^2 = 16$$

In the general form we may write

$$\lim_{x \to a} f(x) = L$$

and we mean that as x approaches a, the limit of f(x) will approach L and L is called the limit of f(x) as x approaches a. No statement is made about f(a) for it may or may not exist although the limit of f(x), as x approaches a, is defined.

If f(x) is defined at

$$x = a$$

and for all values of x near a, and if the function is continuous, then

$$\lim_{x \to a} f(x) = f(a)$$

We are now ready to define a limit.

Let f(x) be defined for all x in the interval near

$$x = a$$

but not necessarily at

$$x = a$$

Then there exists a number L such that for every positive number ϵ

$$\left| f(x) - L \right| < \epsilon$$

provided that we may find the number δ such that

$$0 < |\ x - a\ | < \delta$$

Then we say L is the limit of f(x) as x approaches a and we write

$$\lim_{x \to a} f(x) = L$$

This means that for every challenge number ϵ we must find a number δ in the interval

$$0 < |x - a| < \delta$$

such that the difference between f(x) and L is smaller than the number ϵ.

EXAMPLE: Suppose we are given

$$\lim_{x \to 1} \frac{x^2 + x - 2}{x - 1} = 3$$

and the challenge number ϵ is 0.1.

SOLUTION: We must find a number δ such that in the neighborhood of

$$x = 1$$

for all points except

$$x = 1$$

we have the difference between f(x) and 3 smaller than 0.1.
We write

$$\left| \frac{x^2 + x - 2}{x - 1} - 3 \right| < 0.1$$

and

$$\frac{x^2 + x - 2}{x - 1} - 3$$

$$= \frac{(x + 2)(x - 1)}{x - 1} - 3$$

and we consider only values where

$$x \neq 1$$

Simplifying the first term, we have

$$\frac{(x + 2)(x - 1)}{x - 1} = x + 2$$

Finally, combine terms as follows:

$$x + 2 - 3 = x - 1$$

and

$$|x - 1| < 0.1$$

or

$$-0.1 < x - 1 < 0.1$$

then

$$0.9 < x < 1.1$$

and we have fulfilled the definition of the limit.

If the limit of a function exists and the function is continuous then

$$\lim_{x \to a} f(x) = f(a)$$

For instance, in order to find the limit of the function $x^2 - 3x + 2$ as x approaches 3, we substitute 3 for x in the function. Then

$$f(3) = 3^2 - 3(3) + 2$$

$$= 9 - 9 + 2$$

$$= 2$$

Since x is a variable it may assume a value as close to 3 as we wish, and the closer we choose the value of x to 3, the closer f(x) will approach the value 2. Therefore, 2 is called the limit of f(x) as x approaches 3 and we write

$$\lim_{x \to 3} (x^2 - 3x + 2) = 2$$

Practice Problems

Find the limit of each of the following functions.

1. $\lim\limits_{x \to 1} \dfrac{2x^2 - 1}{2x - 1}$

2. $\lim\limits_{x \to 2} (x^2 - 2x + 3)$

3. $\lim\limits_{x \to a} \dfrac{x^2 - a}{a}$

4. $\lim\limits_{t \to 0} (5t^2 - 3t + 2)$

5. $\lim\limits_{E \to 6} \dfrac{E^3 - E}{E - 1}$

6. $\lim\limits_{Z \to 0} \dfrac{Z^2 - 3Z + 2}{Z - 4}$

Answers

1. 1
2. 3
3. a – 1
4. 2
5. 42
6. $-\dfrac{1}{2}$

Indeterminate Forms

Whenever the answer obtained by substitution, in searching for the value of a limit, assumes any of the following forms, another method for finding the correct limit must be used.

$$\frac{0}{0}, \ \frac{\infty}{\infty}, \ (\infty)\, 0, \ 0^{\circ}, \ \infty^{\circ}, \ 1^{\infty}$$

These are called indeterminate forms.

The proper method for evaluating the limit depends on the problem and sometimes calls for a high degree of ingenuity. We will restrict the methods of solution of indeterminate forms to factoring and division of the numerator and denominator by powers of the variable. Later in the study of limits, L'Hospital's rule will be used as a method of solving indeterminate forms.

Sometimes factoring will resolve an indeterminate form.

EXAMPLE: Find the limit of

$$\frac{x^2 - 9}{x - 3} \ \text{as x approaches 3}$$

SOLUTION: By substitution we find

$$\lim_{x \to 3} \frac{x^2 - 9}{x - 3} = \frac{0}{0}$$

which is an indeterminate form and is therefore excluded as a possible limit. We must now search for a method to find the limit. Factoring is attempted and results in

$$\frac{x^2 - 9}{x - 3} = \frac{(x + 3)(x - 3)}{x - 3}$$

$$= x + 3$$

then

$$\lim_{x \to 3} (x + 3) = 6$$

and we have a determinate limit of 6.

Another indeterminate form is often met when we try to find the limit of a function as the independent variable becomes infinite.

EXAMPLE: Find the limit of

$$\frac{x^4 + 2x^3 - 3x^2 + 2x}{3x^4 - 2x^2 + 1}$$

as x becomes infinite.

SOLUTION: If we let x become infinite in the original expression the result will be

$$\lim_{x \to \infty} \frac{x^4 + 2x^3 - 3x^2 + 2x}{3x^4 - 2x^2 + 1} = \frac{\infty}{\infty}$$

which must be excluded as an indeterminate form. However, if we divide both numerator and denominator by x^4, we obtain

$$\lim_{x \to \infty} \frac{1 + \dfrac{2}{x} - \dfrac{3}{x^2} + \dfrac{2}{x^3}}{3 - \dfrac{2}{x^2} + \dfrac{1}{x^4}}$$

56

$$= \frac{1 + 0 - 0 + 0}{3 - 0 + 0}$$

$$= \frac{1}{3}$$

and we have a determinate limit of $\frac{1}{3}$.

Practice Problems

Find the limit of the following:

1. $\lim\limits_{x \to 2} \dfrac{x^2 - 4}{x - 2}$

2. $\lim\limits_{x \to \infty} \dfrac{2x + 3}{7x - 6}$

3. $\lim\limits_{a \to 0} \dfrac{2a^2b - 3ab^2 + 2ab}{5ab - a^3b^2}$

4. $\lim\limits_{x \to 3} \dfrac{x^2 - x - 6}{x - 3}$

5. $\lim\limits_{x \to a} \dfrac{x^4 - a^4}{x - a}$

6. $\lim\limits_{a \to 0} \dfrac{(x - a)^2 - x^2}{a}$

Answers

1. 4

2. $\dfrac{2}{7}$

3. $\dfrac{2 - 3b}{5}$

4. 5

5. $4a^3$

6. $-2x$

Limit Formulas

To obtain results in calculus we will frequently operate with limits. The proofs of theorems shown in this section will not be given as they are quite long and demand considerable discussion.

The theorems will be stated and examples will be given. Assume that we have three simple functions of x.

$$f(x) = u$$
$$g(x) = v$$
$$h(x) = w$$

Further, let these functions have separate limits such that

$$\lim_{x \to a} u = A$$

$$\lim_{x \to a} v = B$$

$$\lim_{x \to a} w = C$$

Theorem 1. The limit of the sum of two functions is equal to the sum of the limits.

$$\lim_{x \to a} [f(x) + g(x)] = A + B$$

$$= \lim_{x \to a} f(x) + \lim_{x \to a} g(x)$$

This theorem may be extended to include any number of functions such as

$$\lim_{x \to a} [f(x) + g(x) + h(x)] = A + B + C$$

$$= \lim_{x \to a} f(x) + \lim_{x \to a} g(x) + \lim_{x \to a} h(x)$$

EXAMPLE: Find the limit of

$$(x - 3)^2 \quad \text{as } x \to 3$$

SOLUTION:

$$\lim_{x \to 3} (x - 3)^2 = \lim_{x \to 3} (x^2 - 6x + 9)$$

$$= \lim_{x \to 3} x^2 - \lim_{x \to 3} 6x + \lim_{x \to 3} 9$$

$$= 9 - 18 + 9$$
$$= 0$$

Theorem 2. The limit of a constant c times a function f(x) is equal to the constant c times the limit of the function.

$$\lim_{x \to a} cf(x) = cA = c \lim_{x \to a} f(x)$$

59

EXAMPLE: Find the limit of

$$2x^2 \text{ as } x \to 3$$

SOLUTION:
$$\lim_{x \to 3} 2x^2 = 2 \lim_{x \to 3} x^2$$
$$= (2)(9)$$
$$= 18$$

Theorem 3. The limit of the product of two functions is equal to the product of their limits.

$$\lim_{x \to a} f(x)\ g(x) = AB$$

$$= \left(\lim_{x \to a} f(x)\right) \left(\lim_{x \to a} g(x)\right)$$

EXAMPLE: Find the limit of

$$(x^2 - x)\ (\sqrt{2x}) \text{ as } x \to 2$$

SOLUTION:

$$\lim_{x \to 2} (x^2 - x)\ (\sqrt{2x}) = AB$$

$$= \left(\lim_{x \to 2} (x^2 - x)\right) \left(\lim_{x \to 2} \sqrt{2x}\right)$$

$$= (4 - 2)\ (\sqrt{4})$$

$$= 4$$

Theorem 4. The limit of the quotient of two functions is equal to the quotient of their limits, provided the limit of the divisor is not equal to zero.

$$\lim_{x \to a} \frac{f(x)}{g(x)} = \frac{A}{B}$$

$$= \frac{\lim\limits_{x \to a} f(x)}{\lim\limits_{x \to a} g(x)}, \text{ if } B \neq 0$$

EXAMPLE: Find the limit of

$$\frac{3x^2 + x - 6}{2x - 5} \text{ as } x \to 3$$

SOLUTION:

$$\lim_{x \to 3} \frac{3x^2 + x - 6}{2x - 5}$$

$$= \frac{\lim\limits_{x \to 3} 3x^2 + x - 6}{\lim\limits_{x \to 3} 2x - 5}$$

$$= 24$$

Practice Problems

Find the limits ofthe following, using the theorem indicated.

1. $x^2 + x + 2$ as $x \to 1$ (Theorem 1)

2. $7(x^2 - 13)$ as $x \to 4$ (Theorem 2)

3. $5x^4$ as $x \to 2$ (Theorem 3)

4. $\dfrac{2x^2 + x - 4}{3x - 7}$ as $x \to 3$ (Theorem 4)

Answers

1. 4

2. 21

3. 80

4. $\dfrac{17}{2}$

Infinitesimals

In chapter 11, the slope of a curve at a given point was found by taking very small increments of Δy and Δx and the slope was said to be equal to $\dfrac{\Delta y}{\Delta x}$. This section will be a continuation of this concept.

Definitions

A variable that approaches 0 as a limit is called an infinitesimal. This may be written as

$$\lim V = 0$$

or

$$V \to 0$$

and means, as recalled from a previous section of this chapter, that the numerical value of V becomes and remains less than any positive challenge number ϵ.

If the

$$\lim V = L$$

then

$$\lim V - L = 0$$

which indicates that the difference between a variable and its limit is an infinitesimal. Conversely, if the difference between a variable and a constant is an infinitesimal, then the variable approaches the constant as a limit.

EXAMPLE: As x becomes increasingly large, is the term $\frac{1}{x^2}$ an infinitesimal?

SOLUTION: By the definition of infinitesimal, $\frac{1}{x^2}$ approaches 0 as x increases in value, $\frac{1}{x^2}$ is an infinitesimal. It does this, and is therefore an infinitesimal.

EXAMPLE: As x approaches 2, is the expression $\frac{x^2 - 4}{x - 2} - 4$ an infinitesimal?

SOLUTION: By the converse of the definition of infinitesimal, if the difference between $\frac{x^2 - 4}{x - 2}$ and 4 approaches 0, as x approaches 2, the expression $\frac{x^2 - 4}{x - 2} - 4$ is an infinitesimal. By direct substitution we find an indeterminate form; therefore we make use of our knowledge of indeterminates, and write

$$\frac{x^2 - 4}{x - 2} = \frac{(x + 2)(x - 2)}{x - 2} = x + 2$$

and

$$\lim_{x \to 2} (x + 2) = 4$$

The difference between 4 and 4 is 0 and the expression $\frac{x^2 - 4}{x - 2} - 4$ is an infinitesimal, as x approaches 2.

SUMS

An infinitesimal is a variable that approaches 0 as a limit. We state that ϵ and δ, in figure 12-2, are infinitesimals because they both approach 0 as shown.

Theorem 1. The algebraic sum of any number of infinitesimals is an infinitesimal.

In figure 12-2, as ϵ and δ approach 0, notice that their sum approaches 0 and by definition this sum is an infinitesimal and the truth of theorem 1 has been shown. This approach may be used for the sum of any number of infinitesimals.

Products

Theorem 2. The product of any number of infinitesimals is an infinitesimal.

In figure 12-3, the product of two infinitesimals, ϵ and δ, is an infinitesimal as shown. The product of any number of infinitesimals is also an infinitesimal by the same approach as shown for two numbers.

Theorem 3. The product of a constant and an infinitesimal is an infinitesimal.

This may be shown, in figure 12-3, by holding either ϵ or δ constant and noticing their product as the variable approaches 0.

ϵ / δ	1	$\frac{1}{4}$	$\frac{1}{16}$	$\frac{1}{64}$	$\frac{1}{256}$	→ 0
1	2	$\frac{5}{4}$	$\frac{17}{16}$	$\frac{65}{64}$	$\frac{257}{256}$	
$\frac{1}{4}$	$\frac{5}{4}$	$\frac{1}{2}$	$\frac{5}{16}$	$\frac{17}{64}$	$\frac{65}{256}$	
$\frac{1}{16}$	$\frac{17}{16}$	$\frac{5}{16}$	$\frac{1}{8}$	$\frac{5}{64}$	$\frac{17}{256}$	
$\frac{1}{64}$	$\frac{65}{64}$	$\frac{17}{64}$	$\frac{5}{64}$	$\frac{1}{32}$	$\frac{5}{256}$	
$\frac{1}{256}$	$\frac{257}{256}$	$\frac{65}{256}$	$\frac{17}{256}$	$\frac{5}{256}$	$\frac{1}{128}$	
↓ 0						0

Figure 12-2.—Sums of infinitesimals.

$\diagdown$ ε f	1	$\frac{1}{4}$	$\frac{1}{16}$	$\frac{1}{64}$	$\frac{1}{256}$	$\rightarrow 0$
1	1	$\frac{1}{4}$	$\frac{1}{16}$	$\frac{1}{64}$	$\frac{1}{256}$	
$\frac{1}{4}$	$\frac{1}{4}$	$\frac{1}{16}$	$\frac{1}{64}$	$\frac{1}{256}$	$\frac{1}{1024}$	
$\frac{1}{16}$	$\frac{1}{16}$	$\frac{1}{64}$	$\frac{1}{256}$	$\frac{1}{1024}$	$\frac{1}{4096}$	
$\frac{1}{64}$	$\frac{1}{64}$	$\frac{1}{256}$	$\frac{1}{1024}$	$\frac{1}{4096}$	$\frac{1}{16384}$	
$\frac{1}{256}$	$\frac{1}{256}$	$\frac{1}{1024}$	$\frac{1}{4096}$	$\frac{1}{16384}$	$\frac{1}{65536}$	
$\downarrow$ 0						$\searrow$ 0

Figure 12-3.—Products of infinitesimals.

Conclusions

The term infinitesimal was used to describe the term Δx as it approaches zero. The quantity Δx was called in increment of x where increment was used to imply that we have added a small amount to x. Thus $x + \Delta x$ indicates that we are holding x constant and adding a small but variable amount to which we will call Δx.

A very small increment is sometimes called a differential. A small Δx is indicated by dx. The differential of θ is $d\theta$ and that of y is dy. The limit of Δx as it approaches zero is of course zero, but this does not mean that the ratio of two infinitesimals cannot be a real number or a real function of x. For instance, no matter how small Δx is chosen, the ratio $\frac{dx}{dx}$ will still be equal to 1.

In the section on indeterminate forms, a method for evaluating the form $\frac{0}{0}$ was shown. This form results whenever the limit takes the form of one infinitesimal over another. In every case the limit was a real number.

Discontinuities

The discussion of discontinuities will be based upon a comparison to continuity which is defined by:

A function f(x) is continuous at

$$x = a$$

if f(x) is defined at

$$x = a$$

and has a limit as x→a, as follows:

$$\lim_{x \to a} f(x) = f(a)$$

Notice that for continuity a function must fulfil the following three conditions:

1. f(x) is defined at x = a.
2. The limit of f(x) exists as x approaches a.
3. The value of f(x) at x = a is equal to the limit of f(x) at x = a.

If a function f(x) is not continuous at

$$x = a$$

then it is said to be discontinuous at

$$x = a$$

We will use examples to show the above statements.

EXAMPLE: In figure 12-4, is the function

$$f(x) = x^2 + x - 4$$

continuous at f(2)?

SOLUTION:

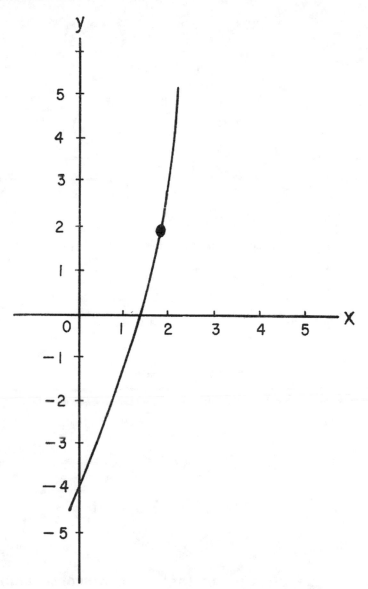

Figure 12-4. -- Function $f(x) = x^2 + x - 4$.

$$f(2) = 4 + 2 - 4$$

$$= 2$$

and

$$\lim_{x \to 2} x^2 + x - 4 = 2$$

and

$$\lim_{x \to 2} f(x) = f(2)$$

Therefore the curve is continuous at

$$x = 2$$

EXAMPLE: In figure 12-5, is the function

$$f(x) = \frac{x^2 - 4}{x - 2}$$

continuous at $f(2)$?

SOLUTION:

$f(2)$ is undefined at

$$x = 2$$

and the function is therefore discontinuous at

$$x = 2$$

However, by extending the original definition of f(x) to read

$$f_1(x) = \begin{cases} \dfrac{x^2 - 4}{x - 2}, & x \neq 2 \\ \\ 4, & x = 2 \end{cases}$$

we will have a continuous function at

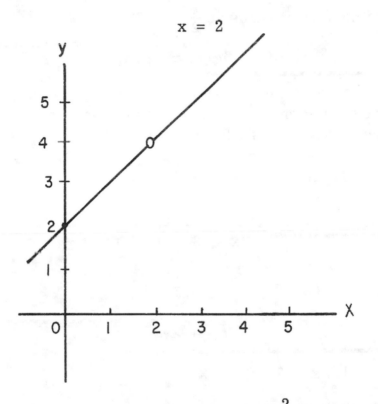

$x = 2$

Figure 12-5.—Function $f(x) = \dfrac{x^2 - 4}{x - 2}$.

NOTE: The value of 4 at x = 2 was found by factoring the numerator of f(x) and then simplifying.

A common kind of discontinuity occurs when dealing with the tangent function of an angle. Figure 12-6 is the graph of the tangent as the angle varies from 0° to 90°; that is, from 0 to $\frac{\pi}{2}$. It should be obvious that the value of the tangent at $\frac{\pi}{2}$ is undefined. Thus the function is said to be discontinuous at $\frac{\pi}{2}$.

Practice Problems

In the folowing definitions of the functions find where the function are discontinuous and then extend the definitions so that the functions are continuous.

1. $f(x) = \dfrac{x^2 - x - 2}{x - 2}$

2. $f(x) = \dfrac{x^2 + 2x - 3}{x + 3}$

3. $f(x) = \dfrac{x^2 + x - 12}{3x - 9}$

Answers

1. x = 2, $f(2) = 3$

2. x = -3, $f(-3) = -4$

3. x = 3, $f(3) = \dfrac{7}{3}$

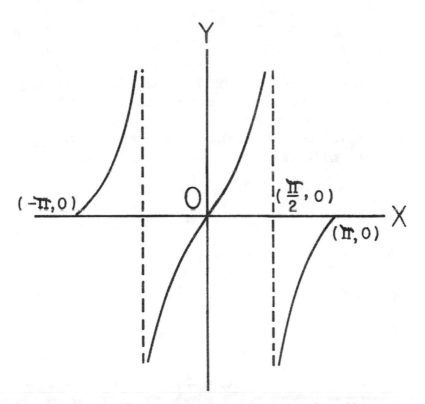

Figure 12-6.—Graph of tangent function.

Increments and Differentiations

In this section we will extend our discussion of limits and examine the idea of the derivative, the heart of differential calculus.

We will assume we have a particular function of x, such that

73

$$y = x^2$$

If x is assigned the value 10, the corresponding value of y will be $(10)^2$ or 100. Now, if we increase the value of x by 2, making it 12, we may call this increase of 2 an increment or Δx. This results in an increase in the value of y and we may call this increase an increment or Δy. From this we write

$$y + \Delta y = (x + \Delta x)^2$$

$$= (10 + 2)^2$$

$$= 144$$

As x increases from 10 to 12, y increases from 100 to 144 so that

$$\Delta x = 2$$

$$\Delta y = 44$$

and

$$\frac{\Delta y}{\Delta x} = \frac{44}{2} = 22$$

We are interested in the ratio $\frac{\Delta y}{\Delta x}$ because the limit of this ratio as Δx approaches zero is the derivative of

$$y = f(x)$$

As recalled from the discussion of limits, as Δx is made smaller, Δy gets smaller also, but the radio $\frac{\Delta y}{\Delta x}$ approaches 20. This is shown in table 12-1.

Table 12-1.—Slope values.

Variable	Values of the variable						
Δx	2	1	0.5	0.2	0.1	0.01	0.0001
Δy	44	21	10.25	4.04	2.01	0.2001	0.00200001
$\frac{\Delta y}{\Delta x}$	22	21	20.5	20.2	20.1	20.01	20.0001

There is a much simpler way to find that the limit of $\frac{\Delta y}{\Delta x}$ as Δx approaches zero is, in this case, equal to 20. We have two equations

$$y + \Delta y = (x + \Delta x)^2$$

and

$$y = x^2$$

By expanding the first equation so that

$$y + \Delta y = x^2 + 2x\,\Delta x + (\Delta x)^2$$

75

and subtracting the second from this, we have

$$\Delta y = 2x \, \Delta x + (\Delta x)^2$$

Dividing both sides of the equation by Δx gives

$$\frac{\Delta y}{\Delta x} = 2x + \Delta x$$

Now, taking the limit as Δx approaches zero

$$\lim_{\Delta x \to 0} \frac{\Delta y}{\Delta x} = 2x$$

Thus,

$$\frac{dy}{dx} = 2x \qquad (2)$$

NOTE: Equation (2) is one way of expressing the derivative of y with respect to x. Other ways are

$$\frac{dy}{dx} = y' = f'(x) = D(x) = \lim_{\Delta x \to 0} \frac{\Delta y}{\Delta x}$$

Equation (2) has the advantage that it is exact and true for all values of x. Thus if

$$x = 10$$

then

$$\frac{dy}{dx} = 2(10) = 20$$

76

and if

$$x = 3$$

then

$$\frac{dy}{dx} = 2(3) = 6$$

This method for obtaining the derivative of y with respect to x is general and may be formulated as follows:

1. Set up the function of x as a function of $(x + \Delta x)$ and expand this function.

2. Subtract the original function of x from the new function $(x + \Delta x)$.

3. Divide both sides of the equation of Δx.

4. Take the limit of all the terms in the equation as Δx approaches zero. The resulting equation is the derivative of f(x) with respect to x.

General Formula

In order to obtain a formula for the derivative of any expression in x, assume the function

$$y = f(x) \tag{3}$$

so that

$$y + \Delta y = f(x + \Delta x) \tag{4}$$

Subtracting equation (3) from equation (4) gives

$$\Delta y = f(x + \Delta x) - f(x)$$

and dividing both sides of the equation by Δx we have

$$\frac{\Delta y}{\Delta x} = \frac{f(x + \Delta x) - f(x)}{\Delta x}$$

The desired formula is obtained by taking the limit of both sides as Δx approaches zero, so that

$$\lim_{\Delta x \to 0} \frac{\Delta y}{\Delta x} = \frac{f(x + \Delta x) - f(x)}{\Delta x}$$

or

$$\frac{dy}{dx} = \lim_{\Delta x \to 0} \frac{f(x + \Delta x) - f(x)}{\Delta x}$$

NOTE: The notation $\frac{dy}{dx}$ is not to be considered as a fraction which has dy for the numerator and dx for the denominator. The expression $\frac{\Delta y}{\Delta x}$ is a fraction with Δy as its numerator and Δx as its denominator and $\frac{dy}{dx}$ is a symbol representing the limit approached by $\frac{\Delta y}{\Delta x}$ as Δx approaches zero.

Examples of Differentiation

In this last section of the chapter we will use several examples of differentiation to obtain a firm understanding of the general formula.

EXAMPLE: Find the derivative $\frac{dy}{dx}$ for the function

$$y = 5x^3 - 3x + 2$$

and determine the slope of its graph at

$$x = -1, \ -\frac{1}{\sqrt{5}}, \ 0, \frac{1}{\sqrt{5}}, \ 1$$

Draw the graph of the function, as shown in figure 12-7

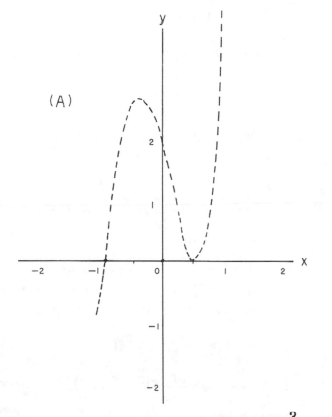

Figure 12-7.—(A) Graph of $f(x) = 5x^3 - 3x$

(B)

x	−1	$-\dfrac{1}{\sqrt{5}}$	0	$\dfrac{1}{\sqrt{5}}$	1
$\dfrac{dy}{dx}$	12	0	−3	0	12
y	0	2.89	2	1.1	4

Figure 12-7.— + 2; (B) chart of values.

SOLUTION: Finding the derivative by formula, we have

$$f(x + \Delta x) = 5(x + \Delta x)^3 - 3(x + \Delta x) + 2 \tag{5}$$

and

$$f(x) = 5x^3 - 3x + 2 \tag{6}$$

Expand equation (5), then subtract equation (6) from equation (5) and simplify to obtain

$$f(x + \Delta x) - f(x)$$

$$= 5\left[3x^2 \Delta x + 3x(\Delta x)^2 + (\Delta x)^3\right] - 3\Delta x$$

Divide through by Δx and we have

80

$$\frac{f(x + \Delta x) - f(x)}{\Delta x}$$

$$= 5 \left[3x^2 + 3x\Delta x + (\Delta x)^2 \right] - 3$$

Take the limit of both sides as $\Delta x \to 0$ and

$$\lim_{\Delta x \to 0} \frac{f(x + \Delta x) - f(x)}{\Delta x} = 15x^2 - 3$$

then

$$\frac{dy}{dx} = 15x^2 - 3$$

Using this derivative let us find the slope of the curve at the points given.

Thus we have a new method of graphing an equation. By substituting different values of x in equation (7) we can find the slope of the curve at the point corresponding to the value of x.

EXAMPLE: Differentiate the function, that is, find $\frac{dy}{dx}$ of

$$y = \frac{1}{x}$$

and then find the slope of the curve at

$$x = 2$$

SOLUTION: Apply the formula for the de-rivative, and simplify as follows:

$$\frac{f(x + \Delta x) - f(x)}{\Delta x} = \frac{\frac{1}{x + \Delta x} - \frac{1}{x}}{\Delta x}$$

$$= \frac{\frac{x - (x + \Delta x)}{x(x + \Delta x)}}{\Delta x}$$

$$= \frac{-1}{x(x + \Delta x)}$$

Now take the limit of both sides as $\Delta x \rightarrow 0$ and

$$\frac{dy}{dx} = \frac{-1}{x^2}$$

In order to find the slope of the curve at the point where x has the value 2, substitute 2 for x in the expression for $\frac{dy}{dx}$:

$$\frac{dy}{dx} = -\frac{1}{2^2}$$

$$= -\frac{1}{4}$$

EXAMPLE: Find the slope of the tangent line on the curve

$$f(x) = x^2 + 4$$

at

$$x = 3$$

SOLUTION: We need to find $\frac{dy}{dx}$ which is the slope of the tangent line at a given point. Apply the formula for the derivative; then,

$$f(x + \Delta x) = (x + \Delta x)^2 + 4 \qquad (8)$$

and

$$f(x) = x^2 + 4 \qquad (9)$$

Expand equation (8) so that

$$f(x + \Delta x) = x^2 + 2x\Delta x + (\Delta x)^2 + 4$$

then subtract equation (9) from equation (8) and

$$f(x + \Delta x) - f(x) = 2x\Delta x + (\Delta x)^2$$

Now, divide through by Δx and

$$\frac{f(x + \Delta x) - f(x)}{\Delta x} = 2x + \Delta x$$

then take the limit of both sides as $\Delta x \to 0$ and

$$\frac{dy}{dx} = 2x$$

Substitute 3 for x in the expression for the derivative to find the slope of the function at

$$x = 3$$

83

so that

$$slope = 6$$

In this last example we will set the derivative of the function f(x) equal to zero to find a maximum or minimum point on the curve. By maximum or minimum of a curve we mean the point or points through which the slope of the curve changes from positive to negative or from negative or positive.

NOTE: When the derivative of a function is set equal to zero this does not mean that in all cases we will have found a maximum or minimum point on the curve. A complete discussion of maxima or minima may be found in most calculus texts.

We will require that the following conditions are met:

1. We have a maximum or minimum point.
2. The derivative exists.
3. We are dealing with an interior point on the curve.

When these conditions are met the derivative of the function will be equal to zero.

EXAMPLE: Find the derivative of the function

$$y = 5x^3 - 6x^2 - 3x + 3$$

and set the derivative equal to zero and find the points of maximum and minimum on the curve, then verify this by drawing the graph of the curve.

SOLUTION: Apply the formula for $\frac{dy}{dx}$ as follows:

$$f(x + \Delta x) =$$

$$\text{(10)}$$

$$5(x + \Delta x)^3 - 6(x + \Delta x)^2 - 3(x + \Delta x) + 3$$

and

$$f(x) - 5x^3 - 6x^2 - 3x + 3 \qquad \text{(11)}$$

Expand equation (10) and subtract equation (11), obtaining

$$f(x + \Delta x) - f(x) = 5(3x^2 \Delta x + 3x\Delta x^2 + \Delta x^3) -$$

$$6(2x\Delta x + \Delta x^2) - 3\Delta x$$

Now, divide through by Δx and take the limit as $\Delta x \rightarrow 0$, so that

$$\frac{dy}{dx} = 5(3x^2) - 6(2x) - 3$$

$$= 15x^2 - 12x - 3$$

Set $\frac{dy}{dx}$ equal to zero, thus

$$15x^2 - 12x - 3 = 0$$

then

$$3(5x^2 - 4x - 1) = 0$$

and

$$(5x + 1)(x - 1) = 0$$

Set each factor equal to zero and find the points of maximum or minimum are

$$5x = -1$$

$$x = -\frac{1}{5}$$

and

$$x = 1$$

The graph of the function is shown in figure 12-8.

Practice Problems

Differentiate the functions in problem 1 through 3.

1. $f(x) = x^2 - 3$

2. $f(x) = x^2 - 5x$

3. $f(x) = 3x^2 - 2x + 3$

4. Find the slope of the curve

$$y = x^3 - 3x + 2$$

at the points

$$x = -2, \ 0, \ \text{and} \ 3$$

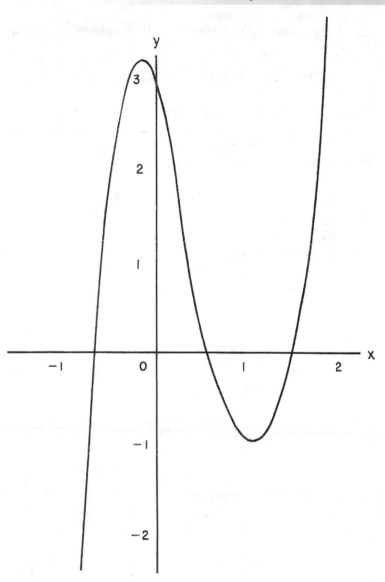

Figure 12-8.—Graph of $5x^3 - 6x^2 - 3x + 3$.

5. Find the values of x where the function

$$f(x) = 2x^3 - 9x^2 - 60x + 12$$

has a maximum or a minimum.

Answers

1. 2x
2. 2x - 5
3. 6x - 2
4. m = 9, -3, and 24
5. x = -2, x = 5

Chapter 13
Derivatives

In the previous chapter on limits, we used the delta process to find the limit of a function as Δx approached zero. We called the result of this tedious and in some cases lengthy process the derivative. In this chapter we will examine some rules used to find the derivative of a function without using the delta process.

To find how y changes as x changes, we take the limit of $\dfrac{\Delta y}{\Delta x}$ as $\Delta x \to 0$ and write

$$\lim_{\Delta x \to 0} \frac{\Delta y}{\Delta x}$$

which is called the derivative of y with respect to x, and we use the symbol $\dfrac{dy}{dx}$ to indicate the derivative and write

$$\lim_{\Delta x \to 0} \frac{\Delta y}{\Delta x} = \frac{dy}{dx}$$

In this section we will take up a number of rules which will enable us to easily obtain the derivative of many algebraic functions. In the derivation of these rules, which will be called theorems, we will assume that

$$\lim_{\Delta x \to 0} \frac{f(x + \Delta x) - f(x)}{\Delta x} = f'(x)$$

or

$$\frac{dy}{dx} = \lim_{\Delta x \to 0} \frac{\Delta y}{\Delta x}$$

exists and is finite.

Derivative of a Constant

The method used to find the derivative of a constant will be similar to the delta process used in the previous chapter but will include an analytical proof. A diagram is used to give a geometrical meaning of the function.

Formula

Theorem 1. The derivative of a constant is zero. Expressed as a formula, this may be written as

$$\frac{dy}{dx} = \lim_{\Delta x \to 0} \frac{\Delta y}{\Delta x} = 0$$

when y is parallel to the x axis.

Proof

In figure 13-1, the graph of

$$y = c \text{ (a constant)}$$

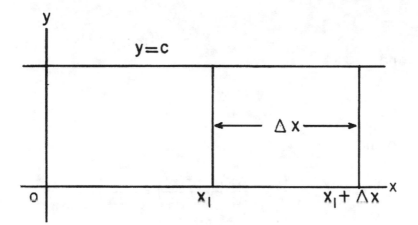

Figure 13-1.—Graph of y = c (a constant).

the value of y is the same for all values of x, and any change in x (that is, Δx) does not affect y, then

$$\Delta y = 0$$

and

$$\frac{\Delta y}{\Delta x} = 0$$

and

$$\frac{dy}{dx} = 0$$

Another way of stating this is that when x is equal to x_1 and when x is equal to $x_1 + \Delta x$, y has the same value. Therefore,

91

$$y = c$$

and

$$y + \Delta y = c$$

so that

$$\frac{\Delta y}{\Delta x} = \frac{f(x + \Delta x) - f(x)}{\Delta x} = \frac{c - c}{\Delta x}$$

and

$$\lim_{\Delta x \to 0} \frac{\Delta y}{\Delta x} = 0$$

then

$$\frac{dy}{dx} = \lim_{\Delta x \to 0} \frac{f(x + \Delta x) - f(x)}{\Delta x}$$

$$= \lim_{\Delta x \to 0} \frac{c - c}{\Delta x} = 0$$

The equation

$$y = c$$

represents a straight line parallel to the x axis. The slope of this line will be zero for all values of x. Therefore, the derivative is zero for all values of x.

EXAMPLE: Find the derivative $\frac{dy}{dx}$ of the function

$$y = 6$$

SOLUTION:

$$y = 6$$

and

$$y + \Delta y = 6$$

therefore

$$\frac{dy}{dx} = \lim_{\Delta x \to 0} \frac{f(x + \Delta x) - f(x)}{\Delta x}$$

$$= \frac{6 - 6}{\Delta x}$$

$$= 0$$

Variables

In this section on variables, we will extend the theorems of limits covered previously. Recall that a derivative is actually a limit. The proof of the theorems presented here involve the delta process, and only a few of these proofs will be offered.

Power Form

Theorem 2. The derivative of the function

$$y = x^n$$

where n is any number is given by

$$\frac{dy}{dx} = nx^{n-1}$$

Proof: By definition

$$\frac{dy}{dx} = \lim_{\Delta x \to 0} \frac{(x + \Delta x)^n - (x)^n}{\Delta x}$$

The expression $(x + \Delta x)^n$ may be expanded by the binomial theorem into

$$x^n + nx^{n-1} \Delta x + \frac{n(n-1)}{2} x^{n-2} \Delta x^2 + \ldots + \Delta x^n$$

Substituting in the expression for the derivative, we have

$$\frac{dy}{dx} = \lim_{x \to 0} \frac{nx^{n-1} \Delta x + \frac{n(n-1)}{2} x^{n-2} \Delta x^2 + \ldots + \Delta x^n}{\Delta x}$$

Simplifying, this becomes

$$\frac{dy}{dx} = \lim_{\Delta x \to 0} \left[nx^{n-1} + \frac{n(n-1)}{2} x^{n-2} \Delta x + \ldots + \Delta x^{n-1} \right]$$

Letting Δx approach zero, we have

$$\frac{dy}{dx} = nx^{n-1}$$

Thus, the proof is complete.

EXAMPLE: Find the derivative of

$$f(x) = x^5$$

SOLUTION: Apply Theorem 2 and find

$$x^5 = x^n$$

therefore

$$n = 5$$

and

$$n - 1 = 4$$

so that given

$$\frac{dy}{dx} = nx^{n-1}$$

and substituting values for n find that

$$\frac{dy}{dx} = 5x^4$$

EXAMPLE: Find the derivative of

$$f(x) = x$$

SOLUTION: Apply Theorem 2 and find

$$x^n = x$$

95

therefore

$$n = 1$$

and

$$n - 1 = 0$$

so that

$$\frac{dy}{dx} = x^{\circ}$$

$$= 1$$

The previous example is a special case of the power form and indicates that the derivative of a function with respect to itself is 1.

EXAMPLE: Find the derivative of

$$f(x) = ax, a = constant$$

SOLUTION:

$$f(x) = ax$$

and

$$f(x + \Delta x) = a(x + \Delta x)$$

$$= ax + a\Delta x$$

so that

$$\Delta y = f(x + \Delta x) - f(x)$$

$$= (ax + a\Delta x) - ax$$

$$= a\Delta x$$

Therefore

$$\frac{dy}{dx} = \lim_{\Delta x \to 0} \frac{a\Delta x}{\Delta x}$$

$$= a$$

The previous example is a continuation of the derivative of a function with respect to itself and indicates that the derivative of a function with respect to itself, times a constant, is that constant.

EXAMPLE: Find the derivative of

$$f(x) = 6x$$

SOLUTION:

$$\frac{dy}{dx} = 6$$

A study of the functions and their derivatives in table 13-1 should further the understanding of this section.

Table 13-1. —Derivatives of functions.

f(x)	3	x	x^2	x^3	x^4	$3x^2$	$9x^3$	x^{-1}	x^{-2}	$3x^{-4}$
$\frac{dy}{dx}$	0	1	2x	$3x^2$	$4x^3$	6x	$27x^2$	$-x^{-2}$	$-2x^{-3}$	$-12x^{-5}$

97

Practice Problems

Find the derivatives of the following:

1. $f(x) = 21$

2. $f(x) = x$

3. $f(x) = 21x$

4. $f(x) = 7x^3$

5. $f(x) = 4x^2$

6. $f(x) = 3x^{-2}$

Answers

1. 0

2. 1

3. 21

4. $21x^2$

5. $8x$

6. $-6x^{-3}$

Sums

Theorem 3. The derivative of the sum of two or more functions of x is equal to the sum of their derivatives.

Assume two functions of x which we will call u and v, such that

$$u = g(x)$$

and

$$v = h(x)$$

and also

$$y = u + v$$
$$= g(x) + h(x)$$

then

$$\frac{dy}{dx} = \frac{du}{dx} + \frac{dv}{dx}$$

Proof:

$$y = g(x) + h(x) \tag{1}$$

$$y + \Delta y = g(x + \Delta x) + h(x + \Delta x) \tag{2}$$

Subtract equation (1) from equation (2) and

$$\Delta y = g(x + \Delta x) + h(x + \Delta x) - g(x) - h(x)$$

Rearrange this equation such that

$$\Delta y = g(x + \Delta x) - g(x) + h(x + \Delta x) - h(x)$$

Divide both sides of the equation by Δx and then take the limit as $\Delta x \to 0$ and

$$\lim_{\Delta x \to 0} \frac{\Delta y}{\Delta x} = \lim_{\Delta x \to 0} \frac{g(x + \Delta x) - g(x)}{\Delta x}$$

$$+ \lim_{\Delta x \to 0} \frac{h(x + \Delta x) - h(x)}{\Delta x}$$

but, by definition

$$\lim_{\Delta x \to 0} \frac{g(x + \Delta x) - g(x)}{\Delta x} = \frac{du}{dx}$$

and

$$\lim_{\Delta x \to 0} \frac{h(x + \Delta x) - h(x)}{\Delta x} = \frac{dv}{dx}$$

then by substitution

$$\frac{dy}{dx} = \frac{du}{dx} + \frac{dv}{dx}$$

EXAMPLE: Find the derivative of the function

$$y = x^3 - 8x^2 + 7x - 5$$

SOLUTION: Theorem 3 indicates that we should find the derivative of each term and then show them as a sum; that is, if

$$y = x^3, \qquad \frac{dy}{dx} = 3x^2$$

$$y = -8x^2, \quad \frac{dy}{dx} = -16x$$

$$y = 7x, \quad \frac{dy}{dx} = 7$$

$$y = -5, \quad \frac{dy}{dx} = 0$$

then, if

$$y = x^3 - 8x^2 + 7x - 5$$

then

$$\frac{dy}{dx} = 3x^2 - 16x + 7 + 0$$

$$= 3x^2 - 16x + 7$$

Practice Problems

Find the derivative of the following:

1. $f(x) = x^2 + x - 1$

2. $f(x) = 2x^4 + 3x + 16$

3. $f(x) = 2x^3 + 3x^2 + x - 3$

4. $f(x) = 3x^3 + 2x^2 - 4x + 2 + 2x^{-1} - 3x^{-3}$

Answers

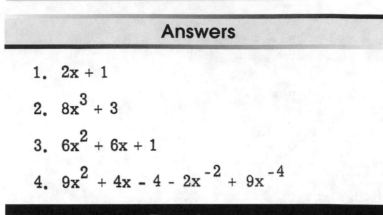

1. $2x + 1$

2. $8x^3 + 3$

3. $6x^2 + 6x + 1$

4. $9x^2 + 4x - 4 - 2x^{-2} + 9x^{-4}$

Products

Theorem 4. The derivative of the product of two functions of x is equal to the first function multiplied by the derivative of the second function, plus the second function multiplied by the derivative of the first function.

If

$$y = uv$$

then

$$\frac{dy}{dx} = u\frac{dv}{dx} + v\frac{du}{dx}$$

This theorem may be extended to include the product of three functions. The result will be as follows:

If

$$y = uvw$$

then

$$\frac{dy}{dx} = uv\frac{dw}{dx} + vw\frac{du}{dx} + uw\frac{dv}{dx}$$

EXAMPLE: Find the derivative of

$$f(x) = (x^2 - 2)(x^4 + 5)$$

SOLUTION: The derivative of the first factor is $2x$, and the derivative of the second factor is $4x^3$. Therefore

$$f'(x) = (x^2 - 2)(4x^3) + (x^4 + 5)(2x)$$

$$= 4x^5 - 8x^3 + 2x^5 + 10x$$

$$= 6x^5 - 8x^3 + 10x$$

EXAMPLE: Find the derivative of

$$f(x) = (x^3 - 3)(x^2 + 2)(x^4 - 5)$$

SOLUTION: The derivatives of the three factors, in the order given, are $3x^2$, $2x$, and $3x^2$.

Therefore

$$f'(x) = (x^3 - 3)(x^2 + 2)(4x^3)$$

$$+ (x^2 + 2)(x^4 - 5)(3x^2)$$

$$+ (x^3 - 3)(x^4 - 5)(2x)$$

103

then

$$f'(x) = 4x^8 + 8x^6 - 12x^5 - 24x^3$$
$$+ 3x^8 + 6x^6 - 15x^4 - 30x^2$$
$$+ 2x^8 - 6x^5 - 10x^4 + 30x$$
$$= 9x^8 + 14x^6 - 18x^5 - 25x^4 - 24x^3 - 30x^2 + 30x$$

Practice Problems

Find the derivatives of the following:

1. $f(x) = x^3(x^2 - 4)$

2. $f(x) = (x^3 - 3)(x^2 + 2x)$

3. $f(x) = (x^2 - 7x)(x^5 - 4x^2)$

4. $f(x) = (x - 2)(x^2 - 3)(x^3 - 4)$

Answers

1. $5x^4 - 12x^2$

2. $5x^4 + 8x^3 - 6x^{-6}$

3. $7x^6 - 42x^5 - 16x^3 + 84x^2$

4. $6x^5 - 10x^4 - 12x^3 + 6x^2 + 16x + 12$

Quotients

Theorem 5. The derivative of the quotient of two functions of x is equal to the denominator times the derivative of the numerator minus the numerator times the derivative of the denominator, all divided by the square of the denominator.

If

$$y = \frac{u}{v}$$

then

$$\frac{dy}{dx} = \frac{v\frac{du}{dx} - u\frac{dv}{dx}}{v^2}$$

EXAMPLE: Find the derivative of the function

$$f(x) = \frac{x^2 - 7}{2x + 8}$$

SOLUTION: The derivative of the numerator is 2x, and the derivative of the denominator is 2. Therefore

$$f'(x) = \frac{(2x + 8)(2x) - (x^2 - 7)(2)}{(2x + 8)^2}$$

$$= \frac{4x^2 + 16x - 2x^2 + 14}{(2x + 8)^2}$$

$$= \frac{2x^2 + 16x + 14}{4(x + 4)^2}$$

$$= \frac{x^2 + 8x + 7}{2(x + 4)^2}$$

Practice Problems

Find the deravitives of the following:

1. $f(x) = \dfrac{x^4}{x^2 - 2}$

2. $f(x) = \dfrac{x^2 - 3}{x + 7}$

3. $f(x) = \dfrac{x^2 + 3x + 5}{x^3 - 4}$

Answers

1. $\dfrac{2x^5 - 8x^3}{(x^2 - 2)^2}$

2. $\dfrac{x^2 + 14x + 3}{(x + 7)^2}$

3. $\dfrac{-(x^4 + 6x^3 + 15x^2 + 8x + 12)}{(x^3 - 4)^2}$

Powers of Functions

Theorem 6. The derivative of any function of x raised to the power n, where n is any number, is equal to n times the polynomial function of x to the (n - 1) power times the derivative of the polynomial itself.

If

$$y = u^n$$

where u is any function of x then

$$\frac{dy}{dx} = nu^{n-1}\frac{du}{dx}$$

EXAMPLE: Find the derivative of the function

$$y = (x^3 - 3x^2 + 2x)^7$$

SOLUTION: Apply Theorem 6 and find

$$\frac{dy}{dx} = 7(x^3 - 3x^2 + 2x)^6 (3x^2 - 6x + 2)$$

EXAMPLE: Find the derivative of the function

$$f(x) = \frac{(x^2 + 2)^3}{x-1}$$

SOLUTION: This problem involves Theorem 5 and Theorem 6. Theorem 6 is used to find the derivative of the numerator, then Theorem 5 is used to find the derivative of the resulting quotient.

107

The derivative of the numerator is

$$3(x^2 + 2)^2 (2x)$$

and the derivative of the denominator is 1. Then, by Theorem 5

$$\frac{dy}{dx} = \frac{(x - 1)[3(x^2 + 2)^2 (2x)] - (1)(x^2 + 2)^3}{(x - 1)^2}$$

$$= \frac{6x(x^2 + 2)^2 (x - 1) - (x^2 + 2)^3}{(x - 1)^2}$$

$$= \frac{(x^2 + 2)^2 [6x(x - 1) - (x^2 + 2)]}{(x - 1)^2}$$

$$= \frac{(x^2 + 2)^2 (5x^2 - 6x - 2)}{(x - 1)^2}$$

Practice Problems

Find the deravitives of the following:

1. $f(x) = (x^3 + 2x - 6)^2$

2. $f(x) = 5(x^2 + x + 7)^4$

3. $f(x) = \dfrac{2(x + 3)^3}{3x}$

Answers

1. $2(x^3 + 2x - 6)(3x^2 + 2)$

2. $20(x^2 + x + 7)^3 (2x + 1)$

3. $\dfrac{18x(x + 3)^2 - 6(x + 3)^3}{9x^2}$

Radicals

To differentiate a function containing a radical, replace the radical by a fractional exponent then find the derivative by applying the appropriate theorems.

EXAMPLE: Find the derivative of

$$f(x) = \sqrt{2x^2 - 5}$$

SOLUTION: Replace the radical by the proper fractional exponent, then

$$f(x) = (2x^2 - 5)^{1/2}$$

and by Theorem 6

$$\frac{dy}{dx} = \frac{1}{2}(2x^2 - 5)^{1/2-1}(4x)$$

$$= \frac{1}{2}(2x^2 - 5)^{-1/2}(4x)$$

$$= 2x(2x^2 - 5)^{-1/2}$$

$$= \frac{2x}{\sqrt{2x^2 - 5}}$$

$$= \frac{2x\sqrt{2x^2 - 5}}{2x^2 - 5}$$

EXAMPLE: Find the derivative of

$$f(x) = \frac{2x + 1}{\sqrt{3x^2 + 2}}$$

SOLUTION: Replace the radical by the proper fractional exponent, thus

$$f(x) = \frac{2x + 1}{(3x^2 + 2)^{1/2}}$$

At this point a decision is in order. This problem may be solved by either writing

$$f(x) = \frac{2x + 1}{(3x^2 + 2)^{1/2}} \tag{3}$$

and applying Theorem 6 in the denominator then applying Theorem 5 for the quotient, or writing

$$f(x) = (2x + 1)(3x^2 + 2)^{-1/2} \tag{4}$$

and applying Theorem 6 for the second factor then applying Theorem 4 for the product.

The two methods of solution will be completed individually as follows:

Use equation (3)

$$f(x) = \frac{2x + 1}{(3x^2 + 2)^{1/2}}$$

Find the derivative of the denominator

$$\frac{d}{dx}(3x^2 + 2)^{1/2}$$

by applying the power theorem and

$$\frac{d}{dx}(3x^2 + 2)^{1/2} = \frac{1}{2}(3x^2 + 2)^{1/2-1}(6x)$$

$$= 3x(3x^2 + 2)^{-1/2}$$

The derivative of the numerator is

$$\frac{d}{dx}(2x + 1) = 2$$

Now apply Theorem 5 and

$$f'(x) = \frac{(3x^2 + 2)^{1/2}(2) - (2x + 1)\left[3x(3x^2 + 2)^{-1/2}\right]}{(3x^2 + 2)}$$

Multiply both numerator and denominator by

$$(3x^2 + 2)^{1/2}$$

and simplify, then

$$f'(x) = \frac{2(3x^2 + 2) - 3x(2x + 1)}{(3x^2 + 2)^{3/2}}$$

$$= \frac{6x^2 + 4 - 6x^2 - 3x}{(3x^2 + 2)^{3/2}}$$

$$= \frac{4 - 3x}{(3x^2 + 2)^{3/2}}$$

To find the same solution, by a different method, use equation (4)

111

$$f(x) = (2x + 1)(3x^2 + 2)^{-1/2}$$

Find the derivative of each factor

$$\frac{d}{dx}(2x + 1) = 2$$

and

$$\frac{d}{dx}(3x^2 + 2)^{-1/2} = -\frac{1}{2}(3x^2 + 2)^{-1/2-1}(6x)$$

$$= -3x(3x^2 + 2)^{-3/2}$$

Now apply Theorem 4 and

$$f'(x) = (2x + 1)\left[-3x(3x^2 + 2)^{-3/2}\right] + (3x^2 + 2)^{-1/2}(2)$$

Multiply both numerator and denominator by

$$(3x^2 + 2)^{-1/2}$$

and

$$f'(x) = \frac{-3x(2x + 1) + 2(3x^2 + 2)}{(3x^2 + 2)^{3/2}}$$

$$= \frac{-6x^2 - 3x + 6x^2 + 4}{(3x^2 + 2)^{3/2}}$$

$$= \frac{4 - 3x}{(3x^2 + 2)^{3/2}}$$

which agrees with the solution of the first method used.

Practice Problems

Find the derivatives of the following:

1. $f(x) = \sqrt{x}$

2. $f(x) = \dfrac{1}{\sqrt{x}}$

3. $f(x) = \sqrt{3x} - 4$

4. $f(x) = 3\sqrt{4x^2 - 3x + 2}$

Answers

1. $\dfrac{1}{2\sqrt{x}}$ or $\dfrac{\sqrt{x}}{2x}$

2. $-\dfrac{1}{2}\sqrt{x^3}$ or $-\dfrac{\sqrt{x^3}}{2x^3}$

3. $\dfrac{3}{2\sqrt{3x} - 4}$ or $\dfrac{3\sqrt{3x} - 4}{2(3x - 4)}$

4. $\dfrac{8x - 3}{3^3\sqrt{(4x^2 - 3x + 2)^2}}$ or

$$\dfrac{(8x - 3)\; 3\sqrt{4x^2 - 3x + 2}}{3(4x^2 - 3x + 2)}$$

Chain Rules

A frequently used rule in differential calculus is the chain rule. This rule links together derivatives which have related variables. The chain rule is

$$\frac{dy}{dx} = \frac{dy}{du}\frac{du}{dx}$$

when the variable y depends on u and u in turn depends on x.

EXAMPLE: Find the derivative of

$$y = (x + x^2)^2$$

SOLUTION: Let

$$u = (x + x^2)$$

and

$$y = u^2$$

Then

$$\frac{dy}{du} = 2u$$

and

$$\frac{dy}{dx} = \frac{dy}{du}\frac{du}{dx} = 2u\frac{du}{dx} \tag{5}$$

Now,

$$\frac{du}{dx} = 1 + 2x$$

and substituting into equation (5) gives

$$\frac{dy}{dx} = \frac{dy}{du}\frac{du}{dx} = 2u(1 + 2x)$$

but

$$u = (x + x^2)$$

therefore,

$$\frac{dy}{dx} = 2(x + x^2)(1 + 2x)$$

EXAMPLE: Find $\frac{dy}{dx}$ where

$$y = 12t^4 + 7t$$

and

$$t = x^2 + 4$$

SOLUTION: By the chain rule

$$\frac{dy}{dx} = \frac{dy}{dt}\frac{dt}{dx}$$

and

$$\frac{dy}{dt} = 48t^3 + 7$$

and

$$\frac{dt}{dx} = 2x$$

then

$$\frac{dy}{dx} = (48t^3 + 7)(2x)$$

and by substitution

$$\frac{dy}{dx} = [48(x^2 + 4)^3 + 7] \ (2x)$$

Practice Problems

Find $\dfrac{dy}{dx}$ in the following:

1. $y = 3t^3 + 8t$ and

   $t = x^3 + 2$

2. $y = 7n^2 + 8n + 3$ and

   $n = 2x^3 + 4x^2 + x$

Answers

1. $[9(x^3 + 2)^2 + 8]\ (3x^2)$

2. $[14(2x^3 + 4x^2 + x) + 8]\ (6x^2 + 8x + 1)$

Inverse Functions

Theorem 7. The derivative of an inverse function is equal to the reciprocal of the derivative of the direct function.

In the equations to this point, x has been the independent variable and y has been the dependent variable. The equations have been in a form such as

$$y = x^2 + 3x + 2$$

Suppose that we have a function like

$$x = \frac{1}{y^2} - \frac{1}{y}$$

and we wish to find the derivative $\frac{dy}{dx}$. Notice that if we solve for y in terms of x, using the quadratic formula, we get the more complicated function

$$y = \frac{-1 \pm \sqrt{1 + 4x}}{2x}$$

If we call this function the direct function, then

117

$$x = \frac{1}{y^2} - \frac{1}{y}$$

is the inverse function. It is easy to determine $\frac{dy}{dx}$ from the inverse function.

EXAMPLE: Find the derivative $\frac{dy}{dx}$ of the function

$$x = \frac{1}{y^2} - \frac{1}{y}$$

SOLUTION: Find the derivative $\frac{dx}{dy}$, thus

$$\frac{dx}{dy} = \frac{-2y}{y^4} + \frac{1}{y^2}$$

$$= \frac{-2}{y^3} + \frac{1}{y^2}$$

$$= \frac{-2 + y}{y^3}$$

The reciprocal of $\frac{dx}{dy}$ is the derivative $\frac{dy}{dx}$ of the direct function, and we find

$$\frac{dy}{dx} = \frac{1}{\frac{dx}{dy}} = \frac{y^3}{y - 2}$$

EXAMPLE: Find the derivative $\frac{dy}{dx}$ of the function

$$x = y^2$$

SOLUTION: Find $\dfrac{dx}{dy}$ to be $\dfrac{d4}{dy} = 2y$

then

$$\frac{dy}{dx} = \frac{1}{\dfrac{dx}{dy}} = \frac{1}{2y}$$

Practice Problems

Find the derivatives $\dfrac{dy}{dx}$ of the functions:

1. $x - 4 - y^2$

2. $x = 9 + y^2$

Answers

1. $-\dfrac{1}{2y}$

2. $\dfrac{1}{2y}$

Implicit Functions

In equations containing x and y, it is not always easy to separate the variables. If we do not solve an equation for y, we call y an implicit function of x. In the equation

$$x^2 - 4y = 0$$

y is an implicit function of x, and x is also called an implicit function of y. If we solved this equation for y, that is

119

$$y = \frac{x^2}{4}$$

then y would be called an explicit function of x. In many cases such a solution would be far too complicated to handle conveniently.

When y is given by an equation such as

$$x^2 + xy^2 = 0$$

y is an implicit function of x.

Whenever we have an equation of this type in which y is a function of x, we can differentiate the function in a straightforward manner. The derivative of each term containing y will be followed by $\frac{dy}{dx}$. Refer to theorem 6.

EXAMPLE: Obtain the derivative $\frac{dy}{dx}$ of the following:

$$x^2 + xy^2 = 0$$

SOLUTION: Find the derivative

$$\frac{d}{dx} (x^2) = 2x$$

and the derivative

$$\frac{d}{dx} (xy^2) = x(2y) \frac{dy}{dx} + (y^2)(1)$$

Therefore,

$$\frac{d}{dx}(x^2 + xy^2) = 2x + 2xy\frac{dy}{dx} + y^2$$

Solving for $\frac{dy}{dx}$ we find that

$$-2xy\frac{dy}{dx} = 2x + y^2$$

and

$$\frac{dy}{dx} = \frac{2x + y^2}{-2xy}$$

$$= -\frac{1}{y} - \frac{y}{2x}$$

Thus, whenever we differentiate an implicit function, the derivative will usually contain terms in both x and y.

Practice Problems

Find the derivative $\frac{dy}{dx}$ of the following:

1. $x^5 + 4xy^3 - 3y^5 = 2$

2. $x^3y^2 = 3$

3. $x^2y + y^3 = 4$

Answers

1. $\dfrac{-5x^4 - 4y^3}{12xy^2 - 15y^4}$

2. $-\dfrac{3y}{2x}$

3. $\dfrac{-2xy}{x^2 + 3y^2}$

Trigonometric Problems

If we are given

$$y = \sin u$$

we may state that, from the general formula,

$$\frac{dy}{du} = \lim_{\Delta u \to 0} \frac{\sin(u + \Delta u) - \sin u}{\Delta u}$$

$$= \lim_{\Delta u \to 0} \frac{\sin u \cos\Delta u + \cos u \sin\Delta u - \sin u}{\Delta u}$$

$$= \lim_{\Delta u \to 0} \frac{\sin u(\cos\Delta u - 1)}{\Delta u} + \lim_{\Delta u \to 0} \frac{\sin\Delta u \cos u}{\Delta u} \tag{6}$$

It can be shown that

$$\lim_{\Delta u \to 0} \frac{\cos\Delta u - 1}{\Delta u} = 0 \tag{7}$$

and

$$\lim_{\Delta u \to 0} \frac{\sin \Delta u}{\Delta u} = 1 \qquad (8)$$

Therefore, by substituting equations (7) and (8) into equation (6)

$$\frac{dy}{du} = \cos u \qquad (9)$$

Now, we are interested in finding the derivative $\frac{dy}{dx}$ of the function $\sin u$ so we apply the chain rule

$$\frac{dy}{dx} = \frac{dy}{du} \frac{du}{dx}$$

and from the chain rule and equation (9) we find

$$\frac{d}{dx} (\sin u) - \cos u \frac{du}{dx}$$

In words, this states that to find the derivative of the sine of a function, we use the cosine of the function times the derivative of the function.

By a similar process we find the derivative of the cosine function to be

$$\frac{d}{dx} (\cos u) = - \sin u \frac{du}{dx}$$

123

The derivatives of the other trigonometric functions may be found by expressing them in terms of the sine and cosine. That is

$$\frac{d}{dx}(\tan u) = \frac{d}{dx}\left(\frac{\sin u}{\cos u}\right)$$

and by substituting sin u for u, cos u for v, and du for dx in the expression of the quotient theorem

$$\frac{d}{dx}\left(\frac{u}{v}\right) = \frac{v\dfrac{du}{dx} - u\dfrac{dv}{dx}}{v^2}$$

we have

$$\frac{dy}{du} = \frac{d}{du}\left(\frac{\sin u}{\cos u}\right)$$

$$= \frac{\cos u \dfrac{d}{du}(\sin u) - \sin u \dfrac{d}{du}(\cos u)}{\cos^2 u} \qquad (10)$$

Taking

$$\frac{d}{du}(\sin u) = \cos u$$

and

$$\frac{d}{du} (\cos u) = - \sin u$$

and substituting into equation (10) find that

$$\frac{dy}{du} = \frac{\cos^2 u + \sin^2 u}{\cos^2 u}$$

$$= \frac{1}{\cos^2 u}$$

$$= \sec^2 u \qquad (11)$$

Now, using the chain rule and equation (11) we find

$$\frac{dy}{dx} = \frac{dy}{du} \frac{du}{dx}$$

$$= \sec^2 u \frac{du}{dx}$$

By stating the other trigonometric functions in terms of the sine and cosine and using similar processes, the following derivatives may be found to be

$$\frac{d}{dx} (\sin u) = \cos u \frac{du}{dx}$$

$$\frac{d}{dx} (\cos u) = - \sin u \frac{du}{dx}$$

125

$$\frac{d}{dx} (\tan u) = \sec^2 u \frac{du}{dx}$$

$$\frac{d}{dx} (\cot u) = -\csc^2 u \frac{du}{dx}$$

$$\frac{d}{dx} (\sec u) = \sec u \tan u \frac{du}{dx}$$

$$\frac{d}{dx} (\csc u) = -\csc u \cot u \frac{du}{dx}$$

EXAMPLE: Find the derivative of the function

$$y = \sin 3x$$

SOLUTION:

$$\frac{dy}{dx} = \cos 3x \frac{d}{dx} (3x)$$

$$= 3 \cos 3x$$

EXAMPLE: Find the derivative of the function

$$y = \tan^2 3x$$

SOLUTION: Use the power theorem and

$$\frac{dy}{dx} = 2 \tan 3x \frac{d}{dx} (\tan 3x)$$

then find

$$\frac{d}{dx}(\tan 3x) = \sec^2 3x \frac{d}{dx}(3x)$$

and

$$\frac{d}{dx}(3x) = 3$$

Combining all of these, we find that

$$\frac{dy}{dx} = (2 \tan 3x)(\sec^2 3x)(3)$$

$$= 6 \tan 3x \sec^2 3x$$

Practice Problems

Find the derivative of the following:

1. $y = \sin 2x$

2. $y = (\cos x^2)^2$

Answers

1. $2 \cos 2x$

2. $-4x \cos x^2 \sin x^2$

Chapter 14
Integration

The two main branches of calculus are differential calculus and integral calculus. Having investigated differential calculus in previous chapters, we now turn our attention to integral calculus. Basically, integration is the inverse of differentiation just as division is the inverse of multiplication, and as subtraction is the inverse of addition.

Definitions

Integration is defined as the inverse of differentiation. When we were dealing with differentiation, we were given a function $F(x)$ and were required to find the derivative of this function. In integration we will be given the derivative of a function and will be required to find the function. That is, when we are given the function $f(x)$, we will find another function $F(x)$ such that

$$\frac{dF(x)}{dx} = f(x) \qquad (1)$$

In words, when we have the function f(x), we must find the function F(x) whose derivative is the function f(x).

If we change equation (1) to read

$$dF(x) = f(x)\ dx \qquad\qquad (2)$$

we have used dx as a differential. An equivalent statement for equation (2) is

$$F(x) = \int f(x)\ dx$$

We call f(x) the integrand, and we say F(x) is equal to the indefinite integral of f(x). The elongated S, that is, $\int$, is the integral sign. This symbol is used because integration may be shown to be the limit of a sum.

Interpretation of an Integral

We will use the area under a curve for the interpretation of an integral. It should be realized, however, that an integral may represent many things, and it may be real or abstract. It may represent plane area, volume, or surface area of some figure.

Area Under a Curve

In order to find the area under a curve, we must agree on what is desired. In figure 14-1, where f(x) is equal to the constant 4, and the "curve" is the straight line

$$y = 4$$

129

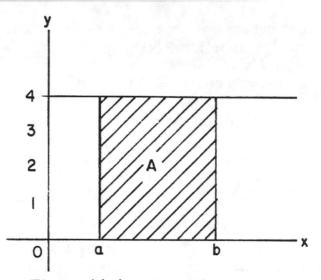

Figure 14-1.—Area of a rectangle.

The area of the rectangle is found by multiplying the height times the width. Thus, the area under the curve is

$$A = 4(b - a)$$

The next problem will be to find a method for determining the area under any curve, provided that the curve is continuous. In figure 14-2, the area under the curve

$$y = f(x)$$

between points x and $x + \Delta x$ is approximately $f(x)\Delta x$. We consider that Δx is small and the area is given to be ΔA. This area under the curve is nearly a rectangle. The area ΔA, under the curve, would differ from the area of the rectangle by the area of the triangle ABC if AC were a straight line.

130

When Δx becomes smaller and smaller, the area of ABC becomes smaller at a faster rate, and ABC finally becomes indistinguishable from a triangle. The area of this triangle becomes negligible when Δx is sufficiently small.

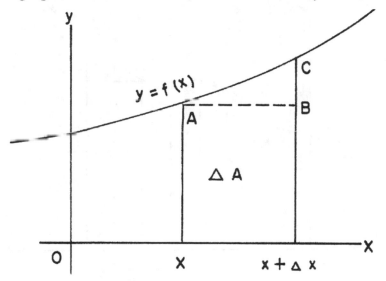

Figure 14-2.—Area ΔA.

Therefore, for sufficiently small values of Δx we can say that

$$\Delta A \approx f(x)\Delta x$$

Now, if we have the curve in figure 14-3, the sum of all the rectangles will be approximately equal to the area under the curve and bounded by the lines at a and b. The difference between the actual area under the curve and the sum of the areas of the rectangles will be the sum of the

131

areas of the triangles above each rectangle.

As Δx is made smaller and smaller, the sum of the rectangular areas will approach the value

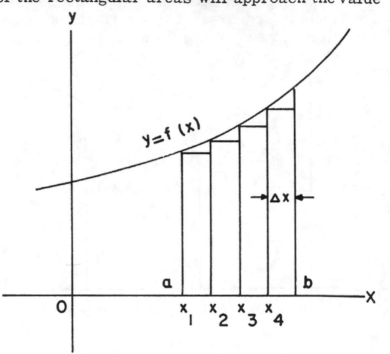

Figure 14-3.—Area of strips.

of the area under the curve. The sum of the areas of the rectangles may be indicated by

$$A = \lim_{n \to \infty} \sum_{k=1}^{n} f(x_k) \, \Delta x \qquad (3)$$

where Σ (sigma) is the symbol for sum, n is the number of rectangles, $f(x) \Delta x$ is the area of each

132

rectangle, and k is the designation number of each rectangle. In the particular example just discussed, where we have four rectangles, we would write

$$A = \sum_{k=1}^{4} f(x_k) \Delta x$$

and we would have only the sum of four rectangles and not the limiting area under the curve.

When using the limit of a sum, as in equation (3), we are required to use extensive algebraic techniques to find the actual area under the curve.

To this point we have been given a choice of using arithmetic and finding only an approximation of the area under a curve, or we could use extensive algebraic preliminaries and find the actual area.

We will now use calculus to find the area under a curve fairly easily.

In figure 14-4, the area under the curve, from a to b, is shown as the sum of the areas of A_{ac} and A_{cb}. The notation A_{ac} means the area under the curve from a to c.

The Intermediate Value Theorem states that

$$_aA_b = f(c) \ (b - a)$$

133

where f(c) in figure 14-4 is the function at an intermediate point between a and b.

We now modify figure 14-4 as shown in figure 14-5.

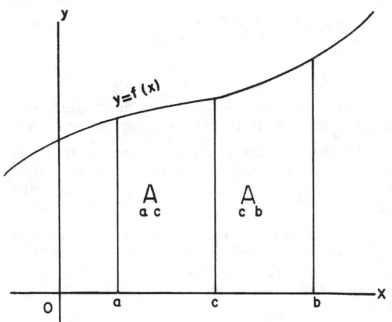

Figure 14-4.—Designation of limits.

When

$$x = a$$

$$_aA_a = 0$$

It is seen in figure 14-5 that

$$_a A_x + {}_x A_{(x+ \Delta x)} = {}_a A_{(x+ \Delta x)}$$

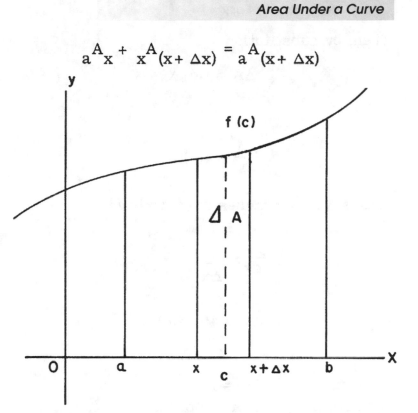

Figure 14-5.—Increments of area at f(c).

therefore, the increase in area, as shown, is

$$\Delta A = {}_x A_{(x+ \Delta x)}$$

but reference to figure 14-5 shows

$${}_x A_{(x+ \Delta x)} = f(c) \Delta x$$

where c is a point between a and b.

135

Then, by substitution

$$\Delta A = f(c) \, \Delta x$$

or

$$\frac{\Delta A}{\Delta x} = f(c)$$

and as Δx approaches zero we have

$$\frac{d(A)}{dx} = \lim_{\Delta x \to 0} \frac{\Delta A}{\Delta x}$$

$$= \lim_{c \to x} f(c)$$

$$= f(x)$$

Now, from the definition of integration

$$_a A_x = \int f(x) \, dx$$

$$= F(x) + C \tag{4}$$

and

$$_a A_a = F(a) + C$$

but

$$_a A_a = 0$$

therefore

$$F(a) + C = 0$$

and solving for C we have

$$C = -F(a)$$

By substituting $-F(a)$ into equation (4) we find

$$_aA_x = F(x) - F(a)$$

If we let

$$x = b$$

then

$$_aA_b = F(b) - F(a) \qquad (5)$$

where $F(b)$ and $F(a)$ are the integrals of the function of the curve at the values b and a.

The constant of integration C is omitted in equation (5) because when the function of the curve at b and a is integrated C will occur with both $F(a)$ and $F(b)$ and will therefore be subtracted from itself.

EXAMPLE: Find the area under the curve

$$y = 2x - 1$$

in figure 14-6, bounded by the vertical lines at a and b, and the x-axis.

SOLUTION: We know that

137

$$_aA_b = F(b) - F(a)$$

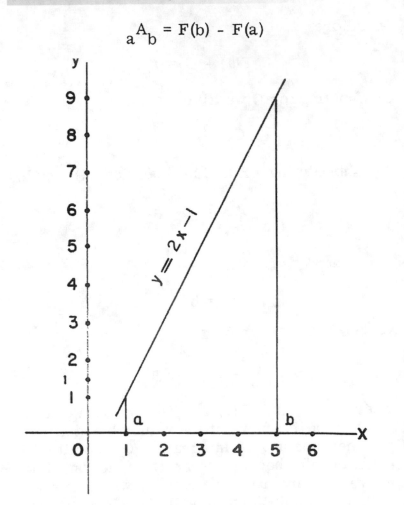

Figure 14-6.—Area of triangle and rectangle.

and find

$$F(x) = \int f(x)dx$$

$$= \int (2x - 1)dx$$

$$= x^2 - x \text{ (this step will be justified later)}$$

Then, substituting the values for a and b into $F(x)$ (that is, $x^2 - x$) find that when

$$x = a$$
$$= 1$$
$$F(a) = 1 - 1$$
$$= 0$$

and when

$$x = b$$
$$= 5$$
$$F(b) = 25 - 5$$
$$= 20$$

Then by substituting these values in

$$_aA_b = F(b) - F(a)$$

find that

$$_aA_b = 20 - 0$$
$$= 20$$

We may verify this by considering figure 14-6 to be a triangle with base 4 and height 8 sitting on a rectangle of height 1 and base 4. By known formulas, we find the area under the curve to be 20.

139

Constant of Integration

A number which is independent of the variable of integration is called a constant of integration. This is to say that two integrals of the same function may differ by the constant of integration.

Integrand

When we are given a differential (or derivative) and we are to find the function whose derivative is the differential we were given, we call the operation integration.

If we have

$$\frac{dy}{dx} = x^2$$

and are asked to find the function whose derivative is this value, x^2, we write

$$dy = x^2 dx$$

then

$$y = \frac{x^3}{3} + C$$

or we write

$$y = \int x^2 dx$$
$$= \frac{x^3}{3} + C$$

The symbol $\int$ is the integral sign, $\int x^2 dx$ is the integral of $x^2 dx$, and x^2 is called the integrand. The C is called the constant of integration.

Indefinite Integrals

When we were finding the derivative of a function, we wrote

$$\frac{dy}{dx} = F(x)$$

or

$$\frac{dF(x)}{dx} = f(x)$$

where we say the derivative of $F(x)$ is $f(x)$. Our problem is to find $F(x)$ when we are given $f(x)$.

We know that the symbol $\int ...dx$ is the inverse of $\frac{d}{dx}$, or when dealing with differentials, the operator symbols d and $\int$ are the inverse of each other.

That is

$$F(x) = \int f(x)dx$$

and when the derivative of each side is taken, d annulling $\int$, we have

$$dF(x) = f(x)dx$$

or where $\int ...dx$ annuls $\frac{d}{dx}$, we have

141

$$\frac{dF(x)}{dx} = \frac{d}{dx} \int f(x)dx$$

$$= f(x)$$

From this, we find that

$$d(x^3) = 3x^2dx$$

then

$$\int 3x^2dx = x^3 + C$$

Also we find that

$$d(x^3 + 3) = 3x^2dx$$

then

$$\int 3x^2dx = x^3 + 2$$

Again, we find that

$$d(x^3 - 9) = 3x^2dx$$

then

$$\int 3x^2dx = x^3 - 9$$

This is to say that

$$d(x^3 + C) = 3x^2dx$$

and

$$3x^2dx = x^3 + C$$

where C is any constant of integration. Since C may have infinitely many values, then a differential expression may have infinitely many integrals which differ only by the constant. We assume the differential expression has at least one integral.

Because the integral contains C and C is indefinite, we call

$$F(x) + C$$

an indefinite integral of $f(x)dx$. In the general form we say

$$\int f(x)dx = F(x) + C$$

With regard to the constant of integration, a theorem and its converse state:

If two functions differ by a constant, they have the same derivative.

If two functions have the same derivative, their difference is a constant.

Evaluating the Constant

To evaluate the constant of integration we will use the following approach.

If we are to find the equation of a curve whose first derivative is 2 times the independent variable x, we may write

$$\frac{dy}{dx} = 2x$$

or

143

$$dy = 2xdx \qquad (6)$$

We may obtain the desired equation for the curve by integrating the expression for dy. That is, integrate both sides of equation (6).
If

$$dy = 2xdx$$

then

$$\int dy = \int 2xdx$$

but

$$\int dy = y$$

and also

$$\int 2xdx = x^2 + C$$

therefore

$$y = x^2 + C$$

We have obtained only a general equation of the curve because a different curve results for each value we assign to C. This is shown in figure 14-7. If we specified that

$$x = 0$$

and

$$y = 6$$

we may obtain a specific value for C and hence a particular curve.

Suppose that

$$y = x^2 + C, \ x = 0, \ \text{and} \ y = 6$$

then

$$6 = 0^2 + C$$

or

$$C = 6$$

By substituting the value 6 into the general equation, the equation for the particular curve is

$$y = x^2 + 6$$

which is curve C of figure 14-7.

The values for x and y will determine the value for C and also determine the particular curve of the family of curves.

In figure 14-7, curve A has a constant equal to -4, curve B has a constant equal to 0, and curve C has a constant equal to 6.

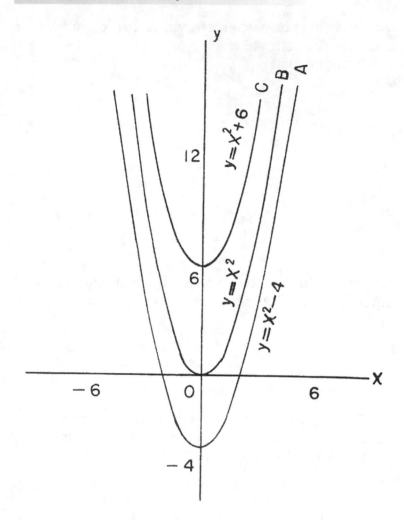

Figure 14-7.—Family of curves.

EXAMPLE: Find the equation of the curve if its first derivative is 6 times the independent variable, y equals 2, and x equals 0.

SOLUTION: We may write

146

$$\frac{dy}{dx} = 6x$$

and

$$\int dy = \int 6xdx$$

therefore

$$y = 3x^2 + C$$

Solving for C when

$$x = 0$$

and

$$y = 2$$

we have

$$2 = 3(0^2) + C$$

or

$$C = 2$$

and the equation is

$$y = 3x^2 + 2$$

Rules for Integration

Although integration is the inverse of differentiation, and we were given rules for dif-

ferentiation, we are required to determine the answers in integration by trial and error. There are some formulas which aid us in the determination of the answer.

In this section we will discuss four of the rules and how they are used to integrate standard elementary forms. In the rules we will let u and v denote a differentiable function of a variable such as x. We will let C, n, and a denote constants.

Our proofs will involve remembering that we are searching for a function $F(x)$ whose derivative is $f(x)dx$.

Rule 1. $\quad\quad \int du = u + C$

The integral of a differential of a function is the function plus a constant.

Proof: If

$$\frac{d(u + C)}{du} = 1$$

then

$$d(u + C) = du$$

and

$$\int du = u + C$$

EXAMPLE: Evaluate the integral

$$\int dx$$

SOLUTION: By Rule 1 we have

$$\int dx = x + C$$

Rule 2. $\int a\,du = a\int du = au + C$

The integral of the product of a constant and a variable is equal to the product of the constant and the integral of the variable. That is, a constant may be moved across the integral sign. NOTE: A variable may NOT be moved across the integral sign.

Proof: If

$$d(au + C) = (a)\, d(u + C)$$

$$= au$$

then

$$\int a\,du = a\int du = au + C$$

EXAMPLE: Evaluate the integral

$$\int 4dx$$

SOLUTION: By Rule 2

$$\int 4dx = 4\int dx$$

and by Rule 1

$$\int dx = x + C$$

therefore

149

$$\int 4dx = 4x + C$$

Rule 3. $\int (du + dv + dw) = \int du + \int dv + \int dw$

$$= u + v + w + C$$

The integral of a sum is equal to the sum of the integrals.

Proof: If

$$d(u + v + w + C) = du + dv + dw$$

then

$$\int (du + dv + dw) = (u + C_1) + (v + C_2)$$

$$+ (w + C_3)$$

$$= u + v + w + C$$

where

$$C = C_1 + C_2 + C_3$$

EXAMPLE: Evaluate the integral

$$\int (2x - 5x + 4)dx$$

SOLUTION: We will not combine 2x and -5x. Then, by Rule 3

$$\int (2x - 5x + 4)dx$$

$$= \int 2xdx - \int + 5xdx + \int 4dx$$

$$= 2 \int xdx - 5 \int xdx + 4 \int dx$$

$$= \frac{2x^2}{2} + C_1 - \frac{5x^2}{2} + C_2 + 4x + C_3$$

$$= x^2 - \frac{5}{2} x^2 + 4x + C$$

where C is the sum of C_1, C_2, and C_3.
This solution requires knowledge of Rule 4 which follows.

Rule 4. $\quad \int u^n du = \frac{u^{n+1}}{n+1} + C$

The integral of $u^n du$ may be obtained by adding 1 to the exponent, then dividing by this new exponent. NOTE: If n is minus 1, this rule is not valid and another method must be used.

Proof: If

$$d\left(\frac{u^{n+1}}{n+1} + C\right) = \frac{(n+1)u^n}{n+1} \ du$$

$$= u^n du$$

then

$$\int u^n du = \frac{u^{n+1}}{n+1} + C$$

EXAMPLE: Evaluate the integral

$$\int x^3 dx$$

SOLUTION: By Rule 4

151

$$\int x^3 dx = \frac{x^{3+1}}{3+1} + C$$

$$= \frac{x^4}{4} + C$$

EXAMPLE: Evaluate the integral

$$\int \frac{7}{x^3} \, dx$$

SOLUTION: First write the integral

$$\int \frac{7}{x^3} \, dx$$

as

$$\int 7x^{-3} dx$$

then, by Rule 2 write

$$7 \int x^{-3} dx$$

and by Rule 4

$$7 \int x^{-3} dx$$

$$= 7 \left(\frac{x^{-2}}{-2} \right) + C$$

$$= -\frac{7}{2x^2} + C$$

EXAMPLE: Evaluate the integral

$$\left(\frac{1}{x^2} + \frac{2}{x^3} \right) dx$$

SOLUTION:

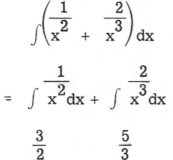

$$\int \left(\frac{1}{x^2} + \frac{2}{x^3} \right) dx$$

$$= \int x^{\frac{1}{x^2}} dx + \int x^{\frac{2}{x^3}} dx$$

$$= \frac{x^{\frac{3}{2}}}{\frac{3}{2}} + C + \frac{x^{\frac{5}{3}}}{\frac{5}{3}} + C$$

$$= \frac{2x^{\frac{3}{2}}}{3} + \frac{3x^{\frac{5}{3}}}{5} + C$$

Practice Problems

Evaluate the following integrals:

1. $\int x^2 dx$

2. $\int 4x \, dx$

3. $\int (x^3 + x^2 + x) dx$

4. $\int 6 \, dx$

5. $\int \frac{5}{x^2} \, dx$

Answers

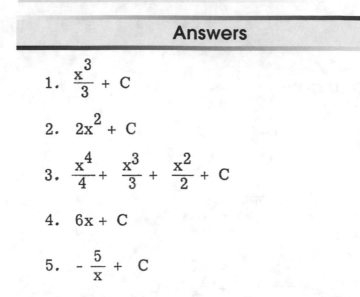

1. $\dfrac{x^3}{3} + C$

2. $2x^2 + C$

3. $\dfrac{x^4}{4} + \dfrac{x^3}{3} + \dfrac{x^2}{2} + C$

4. $6x + C$

5. $-\dfrac{5}{x} + C$

Definite Integrals

The general form of the indefinite integral is

$$\int f(x)dx = F(x) + C$$

and has two identifying characteristics. First, the constant of integration was required to be added to each integration. Second, the result of integration is a function of a variable and has no definite value, even after the constant of integration is determined, until the variable is assigned a numerical value.

The definite integral eliminates these two characteristics. The form of the definite integral is

$$\int_a^b f(x)dx = F(b) + C - [\, F(a) + C \,]$$

$$\tag{7}$$

$$= F(b) - F(a)$$

where a and b are given values. Notice that the constant of integration does not appear in the final expression of equation (7). In words, this equation states that the difference of the values of

$$\int_a^b f(x)dx$$

for

$$x = a$$

and

$$x = b$$

gives the area under the curve defined by f(x), the x axis, and the ordinates where

$$x = a$$

and

$$x = b$$

Upper and Lower Limits

In figure 14-8, the value of b is the upper limit and the value at a is the lower limit. These upper and lower limits may be any assigned values in the range of the curve. The upper limit is positive with respect to the lower limit in that it is located to the right (positive in our case) of the lower limit.

Equation (7) is the limit of the sum of all the strips between a and b, having areas of $f(x)\Delta x$. That is

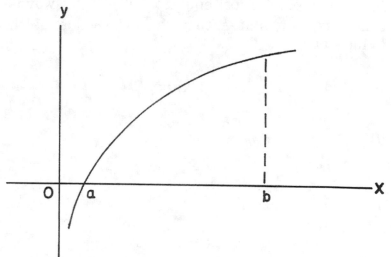

Figure 14-8.—Upper and lower limits.

$$\lim_{\substack{x=a}}^{x=b} \Sigma\ f(x)\Delta x = \int_a^b f(x)dx$$

To evaluate the definite integral

$$\int_a^b f(x)dx$$

find the function $F(x)$ whose derivative is $f(x)dx$

at the value of b and subtract the function at the value of a. That is

$$\int_{a}^{b} f(x)dx = F(x) \Big|_{a}^{b}$$ (8)

$$= F(b) - F(a)$$

where b is the upper limit and a is the lower limit.

EXAMPLE: Find the area bounded by the curve

$$y = x^2$$

the x axis, and the ordinates

$$x = 2$$

and

$$x = 3$$

as shown in figure 14-9.

SOLUTION: Substituting into equation (8)

$$\int_{a}^{b} f(x)dx = F(x) \Big|_{a}^{b} = F(b) - F(a)$$

$$= \int_{2}^{3} x^2 dx$$

$$= \frac{x^3}{3} \Big|_{2}^{3}$$

$$= \frac{3^2}{3} - \frac{2^3}{3}$$

$$= \frac{27}{3} - \frac{8}{3}$$

$$= \frac{19}{3}$$

$$= 6\frac{1}{3}$$

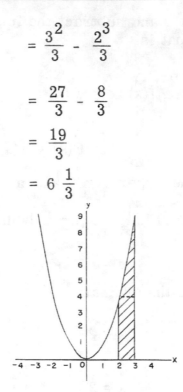

Figure 14-9.—Area from x = 2 to x = 3.

We may make an estimate of this solution by considering the area desired in figure 14-9 as being a right triangle resting on a rectangle The triangle has an approximate area of

$$A = \frac{1}{2}\,bh$$

$$= \frac{1}{2}\,(1)(5)$$

$$= \frac{5}{2}$$

and the area of the rectangle is

$$A = bh$$

$$= (1)(4)$$

$$= 4$$

and

$$4 + \frac{5}{2} = \frac{13}{2} = 6\,\frac{1}{2}$$

which is a close approximation of the area found by the process of integration.

EXAMPLE: Find the area bounded by the curve

$$y = x^2$$

the x axis, and the ordinates

$$x = -2$$

and

$$x = 2$$

as shown in figure 14-10.

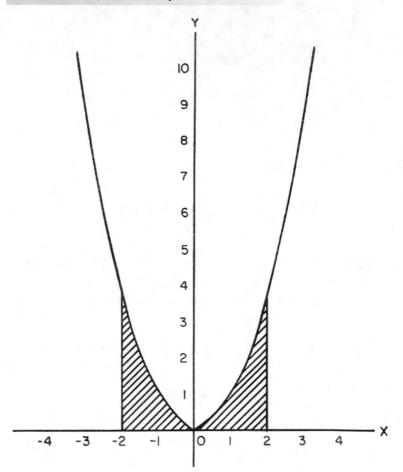

Figure 14-10.—Area under a curve.

SOLUTION: Substituting into equation (8)

$$\int_a^b f(x)dx = F(x) \Big|_{-2}^{2}$$

$$= F(2) - F(-2)$$

160

$$= \int_{-2}^{2} x^2 dx$$

$$= \left. \frac{x^3}{3} \right|_{-2}^{2}$$

$$= \frac{8}{3} - \left[-\frac{8}{3} \right]$$

$$= \frac{16}{3}$$

$$= 5\frac{1}{3}$$

The area above a curve and below the x axis, as shown in figure 14-11, will through integration furnish a negative answer.

Then when dealing with area as shown in figure 14-12, each of the areas shown must be found separately. The areas thus found are then added together, with area considered as the absolute value.

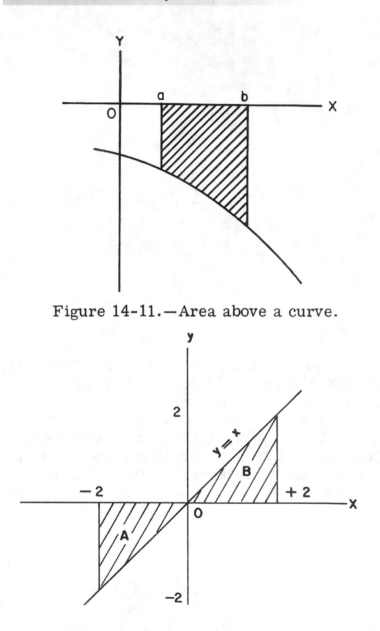

Figure 14-11.—Area above a curve.

Figure 14-12.—Negative and positive value areas

EXAMPLE: Find the area between the curve

$$y = x$$

and the x axis, bounded by the lines

$$x = -2$$

and

$$x = 2$$

SOLUTION: These areas must be computed separately; therefore we write

$$\text{Area A} = \int_{-2}^{0} f(x)dx$$

$$= \int_{-2}^{0} xdx$$

$$= \frac{x^2}{2} \Big|_{-2}^{0}$$

$$= 0 - \left[\frac{4}{2}\right]$$

$$= -2$$

and the absolute value of -2 is

$$|-2| = 2$$

Then

$$\text{Area B} = \int_0^2 f(x)dx$$

$$= \frac{x^2}{2} \Big|_0^2$$

$$= \frac{4}{2} - [0]$$

$$= 2$$

and adding the two areas A and B we find

$$A + B = 2 + 2$$

$$= 4$$

NOTE: INCORRECT SOLUTION: If the function is integrated from -2 to 2 the following incorrect result will occur

$$\text{Area} = \int_{-2}^2 f(x)dx$$

$$= \int_{-2}^2 xdx$$

$$= \frac{x^2}{2} \Big|_{-2}^2$$

$$= \frac{4}{2} - \left[\frac{4}{2}\right]$$

$$= 0 \text{ (INCORRECT SOLUTION)}$$

This is obviously not the area shown in figure 14-12. Such an example emphasizes the value of making a commonsense check on every solution. A sketch of the function will aid this commonsense judgment.

EXAMPLE: Find the total area bounded by the curve

$$y = x^3 - 9x$$

the x axis, and the lines

$$x = -3$$

and

$$x = 3$$

as shown in figure 14-13.

SOLUTION: The area desired is both above and below the x axis; therefore we need to find the areas separately, then add them together using their absolute values.
Therefore

$$A_1 = \int_{-3}^{0} (x^3 - 9x)dx$$

$$= \frac{x^4}{4} - \frac{9}{2} x^2 \Big|_{-3}^{0}$$

$$= 0 - \left[\frac{81}{4} - \frac{81}{2} \right]$$

$$= \frac{81}{4}$$

165

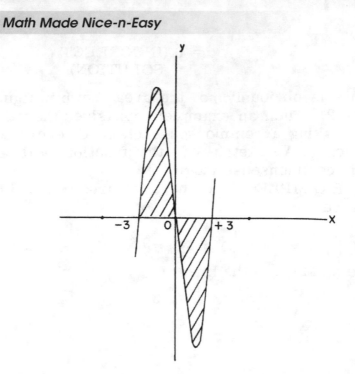

Figure 14-13.—Positive and negative value areas.

The area

$$A_2 = \int_0^3 (x^3 - 9x)dx$$

$$= \frac{x^4}{4} - \frac{9}{2} x^2 \Big|_0^3$$

$$= \frac{81}{4} - \frac{81}{2} - [0]$$

$$= -\frac{81}{4}$$

and

$$\left| -\frac{81}{4} \right| = \frac{81}{4}$$

Then

$$A_1 + A_2 = \frac{81}{4} + \frac{81}{4}$$

$$= \frac{162}{4}$$

$$= 40\frac{1}{2}$$

Practice Problems

1. Find, by integration, the area under the curve

$$y = x + 4$$

bounded by the x axis and the lines

$$x = 2$$

and

$$x = 7$$

Verify this by a geometric process.

2. Find the area under the curve

$$y = 3x^2 + 2$$

bounded by the x axis and the lines

$$x = 0$$

and

$$x = 2$$

3. Find the area between the curve

$$y = x^3 - 12x$$

and the x axis, from

$$x = -1$$

to

$$x = 3$$

Answers

1. $42\frac{1}{2}$

2. 12

3. 39 1/4